STERLING
Education

Essential Chemistry

Electrochemistry

AF703689

4th edition

Copyright © Sterling Education

Customer Satisfaction Guarantee

Your feedback is important because we strive to provide the highest quality educational materials. Email us comments or suggestions.

info@sterling–prep.com

We reply to emails – check your spam folder

All rights reserved. This publication's content, including the text and graphic images, or part thereof, may not be reproduced, downloaded, disseminated, published, converted to electronic media, or distributed by any means whatsoever without the prior written consent from the publisher. Copyright infringement violates federal law and is subject to criminal and civil penalties.

4 3 2 1

ISBN-13: 979-8-8855711-1-1

Sterling Education materials are available at quantity discounts.

Contact info@sterling–prep.com

Sterling Education
6 Liberty Square #11
Boston, MA 02109

© 2026 Sterling Education

Published by Sterling Education

Printed in the U.S.A.

STERLING
Education

From the foundations of chemical reactions to the complex mechanisms of atomic particles, *Essential Chemistry Self-Teaching Guides* are a comprehensive compendium of clearly explained texts to learn and master these multifaceted chemistry topics.

These guides provide a detailed review of the fundamental mechanisms of chemical and physical processes at the atomic level. Develop a better understanding of the electronic structure of elements, principles of chemical bonding, phases of matter, types and mechanisms of chemical reactions, and principles of solutions and acid-base equilibria. Learn about rate processes in chemical reactions, empirical and molecular formulas, enthalpy, entropy, oxidation number, the laws of thermodynamics, and electrochemistry. Reinforce your learning by working through the practice questions and step-by-step solutions.

Created by highly qualified chemistry instructors, researchers, and education specialists, these books empower readers by helping them increase their understanding of general chemistry.

We sincerely hope that these guides are valuable for your learning.

250909akp

Featured on

USA TODAY FOX CBS NBC

Essential Chemistry Self-Teaching Guides

Electronic Structure & Periodic Table

Chemical Bonding

States of Matter & Phase Equilibria

Stoichiometry

Solution Chemistry

Chemical Kinetics & Equilibrium

Acids & Bases

Chemical Thermodynamics

Electrochemistry

Visit our Amazon store

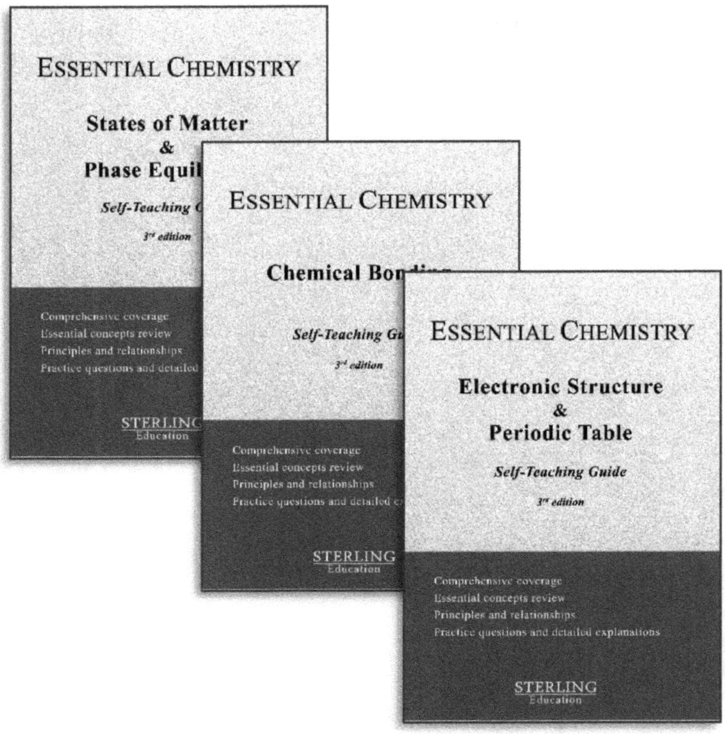

Copyright © Sterling Education.

Essential Physics Self-Teaching Guides

Kinematics and Dynamics

Equilibrium and Momentum

Force, Motion and Gravitation

Work and Energy

Fluids and Solids

Waves and Periodic Motion

Light and Optics

Sound

Electrostatics and Electromagnetism

Electric Circuits

Thermal Physics

Atomic and Nuclear Physics

Visit our Amazon store

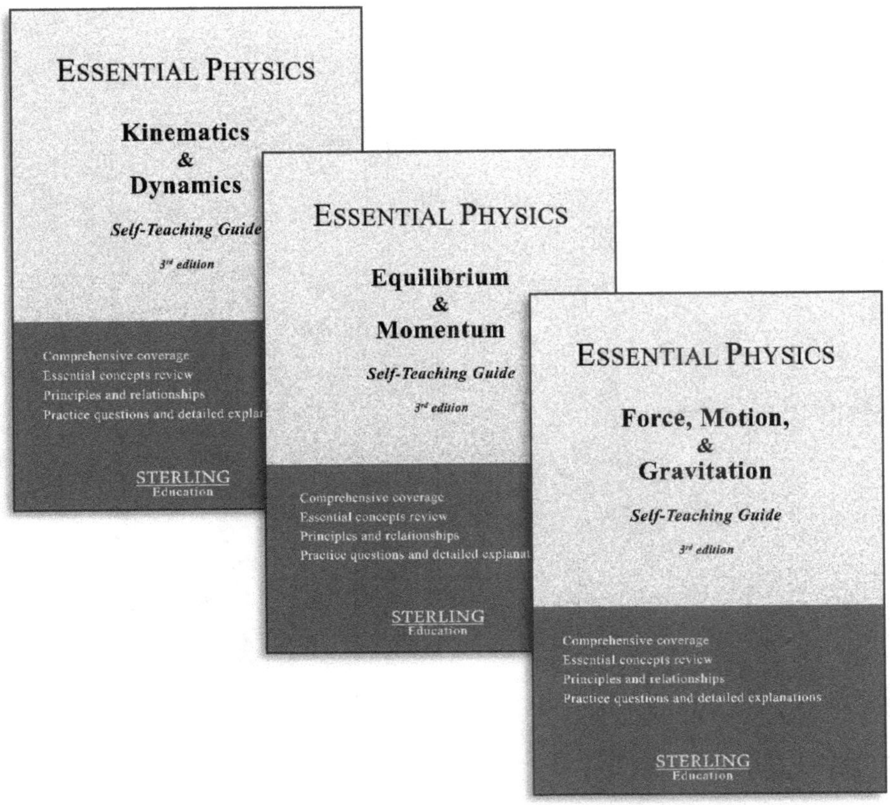

Essential Biology Self-Teaching Guides

Eukaryotic Cell & Cellular Metabolism

Molecular Biology & Genetics

Nervous & Endocrine Systems

Circulatory, Respiratory & Immune Systems

Digestive & Excretory Systems

Muscle, Skeletal & Integumentary Systems

Reproduction & Development

Microbiology

Plants & Photosynthesis

Evolution, Classification & Diversity

Ecology & Population Biology

Visit our Amazon store

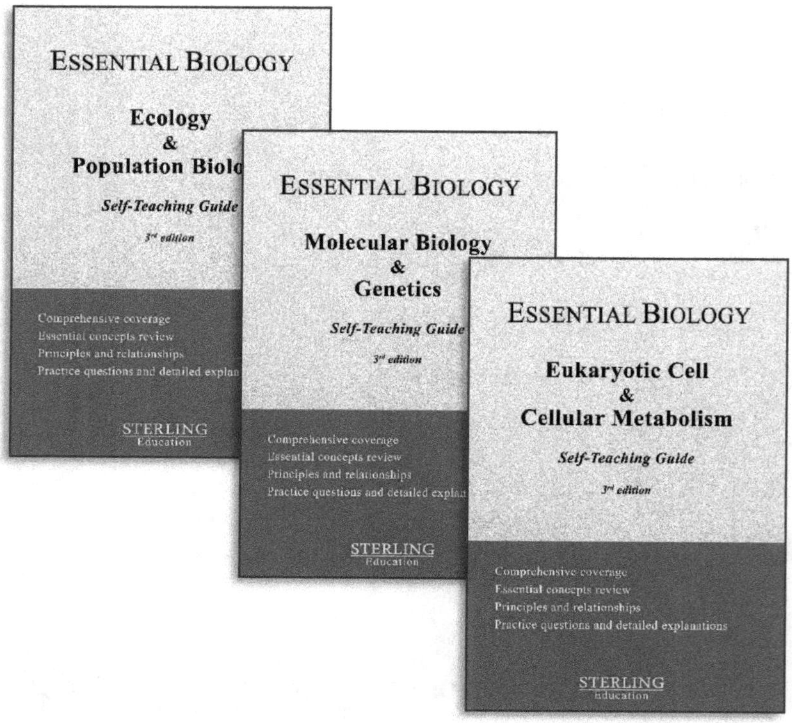

Copyright © Sterling Education.

Everything You Always Wanted to Know About…

Chemistry

Physics

Cell & Molecular Biology

Organismal Biology

Human Anatomy & Physiology

Environmental Science

American History

American Law

American Government & Politics

Comparative Government & Politics

World History

European History

Psychology

Sociology

Human Geography

Visit our Amazon store

For online practice resources visit
https://www.sterling-prep.com

If you benefited from this book, please leave a review on Amazon so others can learn from your input. Reviews help us understand our customers' needs and experiences while keeping our commitment to quality.

Copyright © Sterling Education.

Table of Contents

Any duplication (copies, uploads, PDFs) is illegal.

Table of Contents (*continued*)

Copyright © Sterling Education.

REVIEW

Electrochemistry

Electrochemistry is the study of chemical processes that cause electrons to move. This movement of electrons is called electricity, which can be generated by movements of electrons from one element to another in a reaction known as an oxidation-reduction ("redox") reaction.

Electrochemistry studies the interconversion of electrical and chemical energy. It utilizes spontaneous oxidation-reduction reactions to drive nonspontaneous reactions, converting electrical energy into usable forms. Electrochemistry involves an oxidation-reduction process. Electrochemistry applications include batteries, corrosion control, metallurgy, and electrolysis.

Copyright © Sterling Education.

The Nature of Science

Scientific discovery

Observation is the first step toward scientific discovery. Observations recognize patterns and anomalies in unexplained phenomena.

Hypothesis is advanced as a proposed explanation for the observation.

Evaluating a hypothesis involves experiments and additional observations (i.e., recording and collecting data).

Experimental observations are made while manipulating *independent variables* (e.g., how long the plant is exposed to sunlight per twenty-four hours) with *dependent variables* (data collected regarding height measurements).

Observations generate *data* that:

1) *support the hypothesis* (i.e., different from proves) or

2) *refute the hypothesis*.

Hypothesis are modified by narrowing or expanding their application to explain additional data. More experiments are performed, and the results either *support* (not prove) or *refute* the hypothesis during this iterative process.

Principles are the initially proposed explanations that are specific and apply to a narrow range of phenomena.

Theories build upon valid hypotheses

Theories are proposed to account for the observations when a hypothesis is valid over a range (e.g., sunlight, temperature, or humidity).

Theories are advanced to explain the phenomena and to predict what will happen under other conditions with different variables.

Further observations determine whether the predictions were accurate.

Research (re-search or *again search*) continues as these many observations support hypotheses and theories.

When theories are accurate over various scenarios and experiments, *Laws* are developed based on these repeatedly replicated results.

Any duplication (copies, uploads, PDFs) is illegal.

Laws as established theories

Laws are brief descriptions of how nature behaves in a broad set of circumstances. They are the consolidation of several theories. Laws are the products of multiple theories tested and proven, culminating in a broad claim regarding those predictions.

Laws are widely applicable explanations. Laws with broad and straightforward claims are more robust than those that make claims on a narrower, more specific set of variables and parameters.

The number of Laws describing physical phenomena is minuscule compared to the number of theories and magnitudes, which are less than the number of supported hypotheses.

The number of refuted (unsupported) hypotheses may be several magnitudes larger.

Failed experiments do not typically refer to the physical process where the researcher made a human error during the experiment. A failed experiment means that the *data does not support the hypothesis.* Therefore, after additional trials to validate the "failed" results, the hypothesis must be modified to predict the results of either future experiments or objective observations in the physical world.

Theories, models and laws

Theories emerge from a hypothesis that experimental data has repeatedly supported. Theories are detailed statements that provide testable predictions of the behavior of natural phenomena.

Theories are established to explain observations and then tested (supported or refuted) based on their predictions. Tested theories can articulate the boundaries of the tested hypothesis.

Theories are more specific than *hypotheses* but vaguer and more encompassing than *models.*

Models are constructed to explain the observed behavior if a hypothesis accurately predicts a phenomenon. They are more specific in their application than theories.

Data collected to develop the theories provide the conceptual foundation for constructing models. The model creates mental pictures (e.g., complicated weather patterns displayed as visual pictures for the viewer) of the physical phenomena.

Models should be consistent with scientific theories' predictions and support a comprehensive understanding of a phenomenon. Models allow for predictive outcomes under specific theories (e.g., when the wind speed reaches x, then y will occur).

Multiple models are integrated to explain a portion, or the entire scope, of a *theory.*

Understand the limitations of a model and do not apply it dogmatically unless its applicability is evaluated. A model is an underlying representation of a part of a theory; it is not intended to provide a complete picture of every phenomenon under that theory.

Instead, models provide the fundamental aspects of the theory.

 Copyright © Sterling Education.

Electrochemical Cells

Galvanic and electrolytic cells

Two types of electrochemical cells are *galvanic* (or Voltaic) and *electrolytic.*

Galvanic (or Voltaic) cells convert chemical energy to electrical energy by *spontaneous* redox reactions. A galvanic cell consists of two different metals (electrodes) connected through a conducting solution (an electrolyte) and externally, completing a circuit. A galvanic (voltaic) cell has a resistor or voltmeter in place of the battery.

Galvanic (Voltaic) cells have positive cell potential; thus, no electrical input is required.

Electrolytic cells use electrical energy to drive a non-spontaneous redox reaction, often used to decompose chemical compounds via electrolysis. Electrolytic cells convert electrical energy to chemical energy. Electrolytic cells involve non-spontaneous reactions and thus require an external source of electrons (e.g., a battery).

For example, the electrolysis of sodium chloride (NaCl, common salt) forms sodium (Na^+) metal and chlorine (Cl^-) gas; an electric current supplies the energy required to make the reaction proceed.

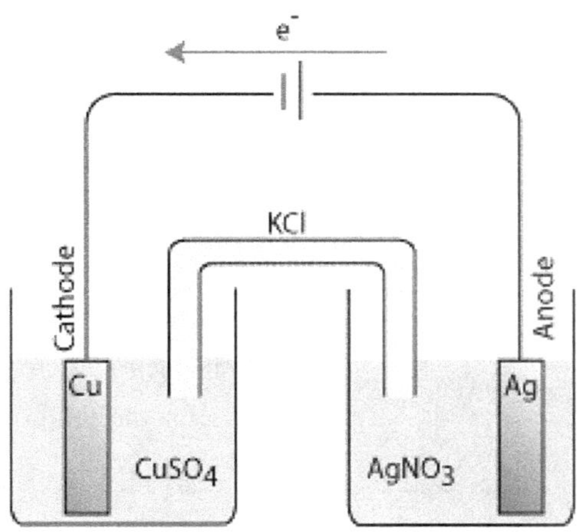

Electrolytic cells use electrical energy to drive non-spontaneous redox reactions

Anode and cathode reactions

Oxidation occurs at the *anode* for an electrochemical cell, and reduction occurs at the *cathode*.

To designate anode and cathode, write a balanced net ionic equation for the spontaneous cell reaction.

- Oxidizing agent (having a more positive or less negative reduction potential, E°) is the cathode.

- Oxidizing agent (less positive or more negative reduction potential E°) is the anode.
- Oxidation occurs in anode half-cell and reduction in cathode half-cell.
- Anode is negative (-), and cathode is positive (+).

Mnemonic: **An Ox** = **AN**ode **Ox**idation **Red Cat** = **RED**uction **CAT**hode

When drawing a galvanic cell, the anode is conventionally placed on the left, with the cathode on the right.

Use the mnemonic **ABC** as a convention (**A**node / **B**ridge / **C**athode).

Galvanic (or Voltaic) cells	**Electrolytic cells**
Cell converts chemical energy into electrical energy.	Cell converts electrical energy into chemical energy.
Redox reaction is *spontaneous* and produces electrical energy.	Redox reaction is *not spontaneous*, and electrical energy must be supplied to initiate the reaction.
Electrons are supplied by the species getting oxidized. The electrons migrate from the anode to the cathode in the external circuit.	An *external battery* supplies the electrons that enter through the cathode and come out through the anode.
Two half-cells are located in different chambers and are connected by a salt bridge.	Both electrodes are placed in the same vessel in the solution.
Anode is *the negative electrode*, and cathode is the positive electrode.	Anode is *the positive electrode*, and cathode is the negative electrode.
Oxidation occurs at the anode, while reduction occurs at the cathode.	Oxidation occurs at the anode, while reduction occurs at the cathode.

Comparison between galvanic and electrolytic cells

Copyright © Sterling Education.

Oxidation and reduction at electrodes

Oxidation occurs at the *anode for electrolytic cells*, while reduction occurs at the cathode.

Electrons originate from the anode because oxidation (loss of electrons) occurs.

$$M \rightarrow M^+ + e^-$$

Electrons travel to the cathode, where they combine with cations on the cathode surface, and reduction (the gain of electrons) occurs.

$$M^+ + e^- \rightarrow M$$

Mnemonic: OIL RIG: **O**xidation **I**s **L**osing e^-, **R**eduction **I**s **G**aining e^-.

Oxidation is an *increase in charge* (more positive) due to the *loss* of e^-.

Reduction is a *decrease in charge* (more negative) due to the *gain* of e^-.

Concentration cells

*Concentration cell*s are electrochemical cells in which the half-cells are of the same type but with different electrolyte concentrations.

The following cell notations are examples of concentration cells:

$$Cu|Cu^{2+}(aq, 0.0010\ M)||Cu^{2+}(aq, 1.0\ M)|Cu$$

$$Ag|Ag^{+}(aq, 0.0010\ M)\ ||Ag^{+}(aq, 0.10\ M)|Ag$$

In concentration cells, the half-cell with the lower electrolyte concentration serves as the anode half-cell, and the other with the higher electrolyte concentration is the cathode half-cell.

At anode half-cell, an oxidation reaction occurs, increasing electrolyte concentration, and at cathode half-cell, a reduction reaction occurs, decreasing the electrolyte concentration.

Any duplication (copies, uploads, PDFs) is illegal.

Notes for active learning

Copyright © Sterling Education.

Electrolysis

Forced current

Electrolysis forces a current through a cell to produce a chemical change, a negative cell potential. The electrolysis process extracts metals from molten salts, and a sustained current is applied to the molten salt.

Electrolysis requires a potential (voltage) input.

Potential (voltage) input + cell potential must be > 0 for reactions to occur.

Electrolytic cells have negative cell potential, so a potential input greater than the cell potential's magnitude is needed for electrolysis.

A battery is a contained unit that produces electricity. Commercial batteries are galvanic cells using solids or pastes as reactants to maximize the electrical output per unit mass.

Fuel cells are a galvanic cell requiring constant reactants to generate electricity.

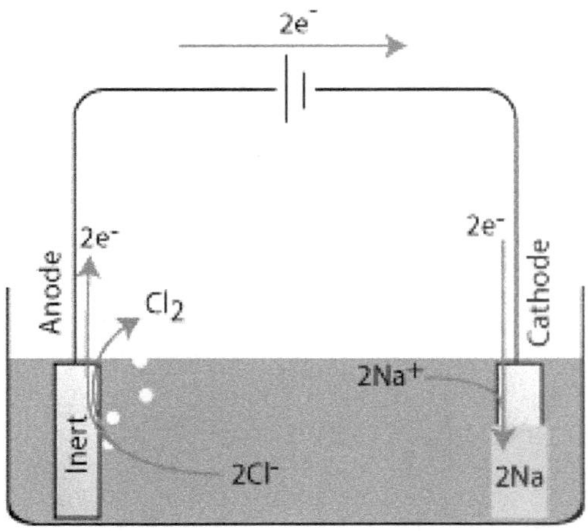

Salt electrolysis charge flow with arrows as how the battery forces electron flow

For example, the following spontaneous reaction occurs in a galvanic (voltaic) cell consisting of $Zn|Zn^{2+}$ (*aq*) and $Cu|Cu^{2+}$ (*aq*) half-cells:

$$Zn\ (s) + Cu^{2+}\ (aq) \rightarrow Zn^{2+}\ (aq) + Cu\ (s); \qquad E^\circ_{cell} = 1.10\ V$$

In an electrolytic cell, the reverse reaction occurs if a voltage greater than the E° cell is applied, which causes the reverse reaction, and excess voltage is *overpotential* or *overvoltage*.

Electrolytic cell reactions

Three types of reactions occur in electrolytic cells if sufficient voltage is applied:

- Solute ions or molecules may be oxidized or reduced.
- Solvent can be oxidized or reduced.
- Metal electrode that forms the anode can be oxidized.

Reactions that occur depend on the *thermodynamic and kinetic properties* of each reaction.

Electric current

Electric current flowing through a cell is measured in *amperes* (A), equal to charge in *coulombs* (C) flowing per second.

1 ampere = 1 coulomb / 1 second

Ampere = Coulomb/second (1 A = 1 C/s)

Coulomb = Ampere × time in sec (1 C = 1 A·s) = total charge

Joule = Coulomb × Volt (1 J = 1 C.V)

Electrolytes are ions that conduct electricity due to their motion. Without electrolytes, electricity cannot propagate.

Electrolysis of water

Decomposition of water is a nonspontaneous process, requiring approximately 400 kJ/mol of energy. The decomposition of water can be affected by electrolysis, where O_2 forms at the *anode* and H_2 at the *cathode.* Assumes an electrolytic cell with:

$[H^+] = [OH^-] = 1$ M and $P_{H2} = P_{O2} = 1$ atm

For electrolysis of pure water, where $[H^+] = [OH^-] = 10^{-7}$ M, the potential for the overall process is –1.23 V.

Anode reaction:	$2\ H_2O \rightarrow \mathbf{O_2} + 4\ H^+(aq) + 4\ e^-$	E° = −1.23 V
Cathode reaction:	$4\ H_2O + 4\ e^- \rightarrow 2\ \mathbf{H_2} + 4\ OH^-(aq)$	E° = −0.83 V
Net reaction:	$6\ H_2O \rightarrow 2\ H_2 + O_2 + 4\ (H^+ + OH^-)$	E° = −2.06 V
	$2\ H_2O \rightarrow 2\ H_2 + O_2$	

A voltage of 1.23 V is enough for the electrolysis of water. However, an additional voltage (overpotential or overvoltage) of approximately 1 V is required to initiate the electrolysis.

 Copyright © Sterling Education.

Electrolysis of solution containing mixtures of ions

Metal with the lowest potential is the first to form.

For example, consider a solution of Cu^{2+}, Zn^{2+} and Ag^{+} electrolyzed with a current with sufficient voltage to reduce the cations.

The reduction potentials of these elements are as follows:

$Ag^{+}(aq) + e^{-} \rightarrow Ag(s)$ $E° = 0.80$ V

$Cu^{2+}(aq) + 2e^{-} \rightarrow Cu(s)$ $E° = 0.34$ V

$Zn^{2+}(aq) + 2e^{-} \rightarrow Zn(s)$ $E° = -0.76$ V

Since the reduction of Ag^{+} to Ag has the most positive potential, Ag deposits at the cathode before other metals, followed by Cu and Zn, respectively. If the voltage is controlled, starting with the lowest voltage, it is possible to separate the three metals using electrolysis.

For example, how many grams of Ag are deposited in the cathode of an electrolytic cell if a current of 3.50 A is applied to a solution of $AgNO_3$ for 12 minutes? (Use the molecular mass of Ag = 107.86 g/mol and the conversion of 1 mol $e^{-} = 9.65 \times 10^{4}$ C)

Balanced equation:

$Ag^{+} + e^{-} \rightarrow Ag(s)$

Formula to calculate deposit mass:

mass of deposit = (atomic mass × current × time) / 96,500 C

mass of deposit = [107.86 g × 3.50 A × (12 min × 60 s/min)] / 96,500 C

mass of deposit = 2.82 g

Stoichiometry of electrolysis

Number of substances formed at the anode or cathode is calculated from the current's magnitude (amperes) and time (sec).

For example, if a current of 1.50 A flows through an aqueous solution of $CuSO_4$ for 25.0 minutes, the amount of *charge passing* through the solution is:

charge = (1.50 C/s)·(25.0 minutes)·(60 sec/min)

charge = 2,250 C

Number of *moles of electrons* passing through the solution:

moles of electrons = (2,250 C) × (1 mol e^{-})

moles of electrons = 0.0233 mol

Any duplication (copies, uploads, PDFs) is illegal.

Reduction of Cu^{2+} requires 2 mol e^- per mole of Cu:

$$Cu^{2+} + 2\ e^- \rightarrow Cu$$

Cu formed at cathode:

$$(0.0233\ \text{mol}\ e^-) \times \frac{(1\ \text{mol Cu})}{(2\ \text{mol}\ e^-)} \times \frac{(63.55\ \text{g Cu})}{(1\ \text{mol Cu})}$$

Cu formed at cathode = 0.741 g Cu

If the process uses a cell with 3.0 V:

energy consumed = 3.0 V × 2250 C

energy consumed = 6,750 C.V

6,750 C.V = 6,800 J

6,800 J = 6.8 kJ

Faraday's Law relates the number of elements

In 1833, Michael Faraday (1791-1867) proposed quantitative laws to express magnitudes of electrolytic effects. The mass of elements deposited at an electrode is directly proportional to the charge (A·sec or Coulombs).

Faraday constant is the electric charge of 1 mole of electrons, a conversion factor between moles and coulombs. Faraday constant (F):

$$F = [6.02 \times 10^{23}\ \text{electrons/mol}\ e^-] / [6.24 \times 10^{18}\ e^-/\text{C}]$$

$$F = 96{,}485\ \text{C/mol}\ e^-$$

1 mol electrons is 6.022×10^{23} electrons, and a coulomb is the (negative) charge of 6.241×10^{18} electrons. The Faraday constant is the quotient of these.

For electrolysis, divide the charge in coulombs by the Faraday constant to calculate the number of moles that underwent electrolysis.

Current I = coulombs of charge per second

$$I = q / t$$

Faraday's constant:

F = coulombs of charge per mol of electron

F = total charge / total moles of electrons

$$F = q / n$$

$q = It$ and $q = \text{nF}$

Copyright © Sterling Education.

Thus:

$$It = nF$$

Current × time = moles of e^- × Faraday's constant.

Using this equation, solve for moles of electrons (n). Using the half-equation stoichiometry, determine how many moles of the element are made for every e^- transferred.

For example, 1 mol of Cu is deposited for every 2 moles of electrons for the half-reaction, using Faraday's constant = 96,485 C/mo.

$$Cu^{2+} + 2\ e^- \rightarrow Cu$$

For example, if 1 amp of current passes a cathode for 10 minutes, how much Zn (s) forms in the following reaction? (Use the molecular mass of Zn = 65 g/mole)

$$Zn^{2+} + 2\ e^- \rightarrow Zn\ (s)$$

Calculate the mass of metal deposited at the cathode:

Step 1: ***Calculate total charge using current and time***

Q = current × time

Q = 1 A × (10 minutes × 60 s/minute)

Q = 600 A·s = 600 C

Step 2: ***Calculate moles of electrons that have the same amount of charge***

moles e^- = Q / 96,500 C/mol

moles e^- = 600 C / 96,500 C/mol

moles e^- = 6.22×10^{-3} mol

Step 3: ***Calculate moles of metal deposit***

Half-reaction of zinc ion reduction:

$$Zn^{2+}\ (aq) + 2\ e^- \rightarrow Zn\ (s)$$

moles of Zn = (coefficient Zn / coefficient e^-) × moles e^-

moles of Zn = (½) × 6.22×10^{-3} mol

moles of Zn = 3.11×10^{-3} mol

Any duplication (copies, uploads, PDFs) is illegal.

Step 4: ***Calculate mass of metal deposit***

mass Zn = moles Zn × molecular mass of Zn

mass Zn = 3.11×10^{-3} mol × (65 g/mol)

mass Zn = 0.20 g

For example, how long would it take to deposit 4.00 grams of Cu from a $CuSO_4$ solution if a current of 2.5 A is applied? (Use molecular mass of Cu = 63.55 g/mol and conversion of 1 mol e^- = 9.65×10^4 C and 1 C = A·s)

Balanced equation:

$Cu^{2+} + 2\ e^- \rightarrow Cu\ (s)$

Formula to calculate deposit mass:

mass of deposit = atomic mass × moles of electron

Since Cu has 2 electrons per atom, multiply the moles by 2:

mass of deposit = atomic mass × (2 × moles of electron)

mass of deposit = atomic mass × (2 × current × time) / 96,500 C

4.00 g = 63.55 g × (2 × 2.50 A × time) / 96,500 C

time = 4,880 s

time = (4,880 s × 1 min/60s × 1 hr/60 min)

time = 1.36 hr

Copyright © Sterling Education.

Galvanic (Voltaic) Cells

Spontaneous oxidation-reduction reactions

Galvanic cells are a typical application of electrochemistry in batteries. A galvanic cell (i.e., *voltaic cell*) uses a spontaneous oxidation-reduction reaction to produce an electric current. Galvanic cells produce voltage and were invented between 1780 and 1799 by Italian physicists Luigi Galvani (1737-1798) and Alessandro Volta (1745-1827).

In galvanic cells, species with a more *positive reduction potential* serve as the *cathode*, and those with a *less positive or more negative reduction potential* serve as the *anode* half-cell.

Galvanic cells:

- *Oxidation* occurs at the *anode.*
- *Reduction* occurs at the *cathode.*
- Salt bridges or porous disks allow ions to flow without extensive mixing of the solutions.

Salt bridge contains a strong electrolyte held in a gel-like matrix.

Porous disk contains tiny passages for the hindered flow of ions.

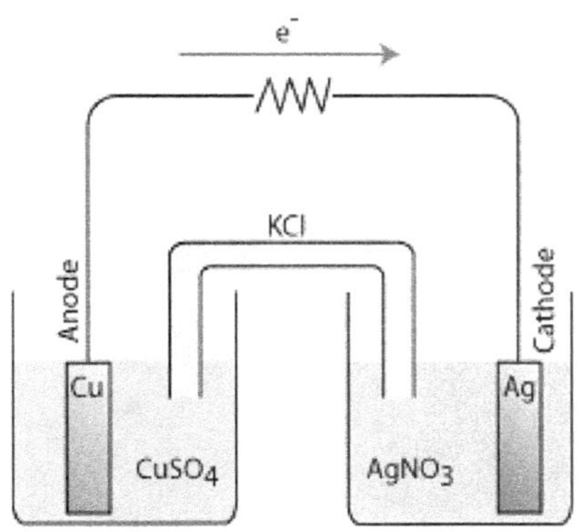

Galvanic (Voltaic) Cell

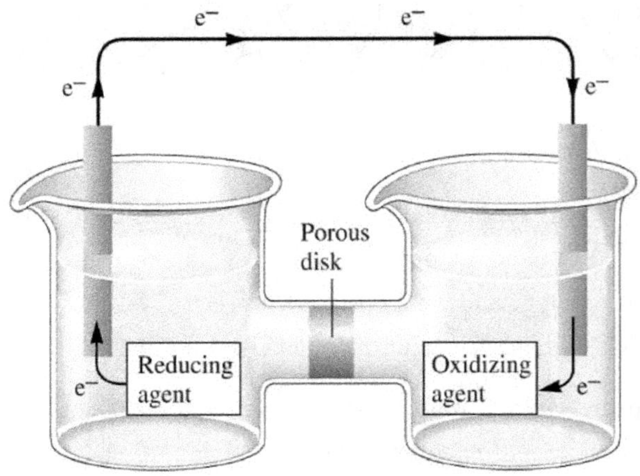

anode *cathode*

Oxidation-reduction reaction examples

For example, spontaneous oxidation-reduction reactions:

$$Zn\,(s) + CuSO_4\,(aq) \rightarrow Cu\,(s) + ZnSO_4\,(aq)$$

$$Zn\,(s) + Cu\cancel{SO_4}\,(aq) \rightarrow Cu\,(s) + Zn\cancel{SO_4}\,(aq)$$

$$Zn\,(s) + Cu^{2+}\,(aq) \rightarrow Cu\,(s) + Zn^{2+}\,(aq)$$

$$Cr\,(s) + 3\,AgNO_3\,(aq) \rightarrow Cr(NO_3)_3\,(aq) + 3\,Ag\,(s)$$

$$Cr\,(s) + 3\,Ag\cancel{NO_3}\,(aq) \rightarrow Cr\cancel{(NO_3)_3}\,(aq) + 3\,Ag\,(s)$$

$$Cr\,(s) + 3\,Ag^{+}\,(aq) \rightarrow Cr^{3+}\,(aq) + 3\,Ag\,(s)$$

$$MnO_4^{-}\,(aq) + 5\,Fe^{2+}\,(aq) + 8\,H^{+}\,(aq) \rightarrow Mn^{2+}\,(aq) + 5\,Fe^{3+}\,(aq) + 4\,H_2O\,(l)$$

$$Zn\,(s) + 2\,MnO_2\,(s) \rightarrow ZnO\,(s) + Mn_2O_3\,(s)$$

Redox half-reactions

Oxidation half-reaction is the species that loses electrons (increasing its charge).

Oxidation example,

$$Cu \rightarrow Cu^{2+} + 2\,e^{-}$$

Reduction half-reaction is the species gaining electrons (decreasing charge).

Reduction example,

$$2\,Ag^{+} + 2\,e^{-} \rightarrow 2\,Ag$$

 Copyright © Sterling Education.

Ten steps for balancing redox half-reactions

1. Split the equation into two half-reactions.
2. Balance atoms other than O or H.
3. Balance O using H_2O. Add water as needed to balance oxygen on side deficient in oxygen.
4. If in an acidic solution, balance hydrogens using H^+.

   Add H^+ ions as needed to the side deficient in H.
5. In basic solution, add OH^- ions equal to the number of H^+ ions to each side of the chemical equation.
6. The mass is balanced. To balance charge, determine the charge on each side of the reaction and add the necessary electrons to the positive side, so that the charges are equal.
7. Repeat steps 1 through 4 for each half-reaction.
8. If electrons lost do not equal the number gained, multiply each half-reaction by the necessary factor.
9. Sum half-reactions to obtain the balanced net ionic reaction. Inspect H^+ ions and H_2O molecules to cancel.
10. Check the final equation for mass and charge balance.

Important: If expressing H^+ as H_3O^+, change the number of H^+ into the same number of H_3O^+.

Add that same number of H_2O to the opposite side of the equation.

Redox reactions can be separated into two half-reactions:

Example 1:

Oxidation half-reaction: $Zn\ (s) \rightarrow Zn^{2+}\ (aq) + 2\ e^-$

Reduction half-reaction: $Cu^{2+}\ (aq) + 2\ e^- \rightarrow Cu\ (s)$

Overall reaction: $Zn\ (s) + Cu^{2+}\ (aq) \rightarrow Cu\ (s) + Zn^{2+}\ (aq)$

Any duplication (copies, uploads, PDFs) is illegal.

Example 2:

Oxidation half-reaction:

$$5\ Fe^{2+}(aq) \rightarrow 5\ Fe^{3+}(aq) + 5\ e^-$$

Reduction half-reaction:

$$MnO_4^-(aq) + 8\ H^+(aq) + 5\ e^- \rightarrow Mn^{2+}(aq) + 4\ H_2O\ (l)$$

Overall reaction:

$$MnO_4^-(aq) + 5\ Fe^{2+}(aq) + 8\ H^+(aq) \rightarrow Mn^{2+}(aq) + 5\ Fe^{3+}(aq) + 4\ H_2O\ (l)$$

Galvanic cell half-reactions

Redox reactions are split into two half-reactions in *galvanic cells*, each occurring in a separate compartment as a half-cell. Chemical energy drives electrons through the external circuit, connecting the two half-cells, and thus produces an electric current.

For example, consider the reaction:

$$Zn\ (s) + CuSO_4\ (aq) \rightarrow Cu\ (s) + ZnSO_4\ (aq)$$

If zinc metal is placed in the $CuSO_4$ solution, an exothermic reaction occurs, producing heat. In a galvanic cell, zinc metal is placed in a $ZnSO_4$ solution in one container, and copper metal is placed in a $CuSO_4$ solution in another container. A wire connects two metals to complete the circuit; the two solutions are connected by a salt bridge containing strong electrolytes, such as $KCl(aq)$ or $K_2SO_4(aq)$, which allow ions to flow between the two half-cells. The reactants – zinc metal and copper ions – are not allowed to come into direct contact.

Due to the potential difference between the two half-cells, electrons are forced to flow from the zinc electrode (the anode) to the copper electrode (the cathode).

At the interface between the zinc electrode and the $ZnSO_4$ solution, zinc atoms are oxidized to Zn2+ ions; at the interface of the copper electrode and the CuSO4 solution, Cu2+ ions combine with incoming electrons, reducing them to Cu atoms.

Half-reaction equations represent these oxidation and reduction processes:

Anode-reaction (oxidation):

$$Zn\ (s) \rightarrow Zn^{2+}(aq) + 2\ e^-$$

Cathode-reaction (reduction):

$$Cu^{2+}(aq) + 2\ e^- \rightarrow Cu\ (s)$$

 Copyright © Sterling Education.

Cell notation

Two half-cells are represented by notations $Zn|Zn^{2+}$ and $Cu^{2+}|Cu$.

In a galvanic cell, the more readily oxidized metal serves as the anode, and the other serves as the cathode.

Determine the species' tendency to oxidize from the standard reduction potentials table.

Zinc is more easily oxidized than copper, serving as the anode, while copper forms the cathode.

Oxidation half-reaction occurs at the anode

Reduction half-reaction at the cathode.

This galvanic cell is represented using the following cell notation:

$Zn|ZnSO_4\,(aq)||CuSO_4\,(aq)|Cu$

Cell notation has the anode on the left side, with the cathode on the right.

At the anode half-cell, Zn^{2+} ions are continuously formed, creating an excess of positive ions.

Cu2+ ions are continuously reduced to Cu at the cathode half-cell, causing a decrease in cation concentration.

To maintain electrically neutral solutions in half-cells, anions flow into the anode half-cell, and cations flow into the cathode half-cell. Thus, an electric current is the flow of charged particles. Electrons flow from anode to cathode through the wire; cations and anions flow in the opposite direction through the salt bridge.

Any duplication (copies, uploads, PDFs) is illegal.

Notes for active learning

Copyright © Sterling Education.

Reduction Potential

Standard reduction potential

Standard half-cell potential (or *reduction potential*) is determined by connecting the half-cell (under standard conditions) to the *standard hydrogen electrode* (SHE) as a reference half-cell.

This reference half-cell consists of a Pt electrode in a solution containing 1 M H^+, into which H_2 gas is purged at a constant pressure of 1 atm.

Since the reference half-cell is assigned zero potential, the cell potential measured under these conditions is the substance's standard half-cell potential.

For example, when a $Zn|Zn^{2+}$(*aq*, 1 M) half-cell is connected to the standard hydrogen electrode (SHE), the voltage measured at 25 °C is found to be 0.763 V.

Electrons flow from $Zn|Zn^{2+}$ to SHE; relative to SHE, the reduction potential for the $Zn|Zn^{2+}$ half-cell is the negative value of the measured voltage. (E° = reduction potential)

$$Zn^{2+}(aq) + 2\,e^- \rightarrow Zn(s) \qquad E° = -076\ V$$

$$2\,H^+(aq) + 2\,e^- \rightarrow H_2(s) \qquad E° = 0.000\ V$$

$$Cu^{2+}(aq) + 2\,e^- \rightarrow Cu(s) \qquad E° = 0.34\ V$$

Since values are obtained against the standard hydrogen potential, species with positive reduction potentials are easier to reduce than H^+ ions.

In contrast, those with negative reduction potentials are more difficult to reduce.

Therefore, relative to H^+, Cu^{2+} is more easily reduced, whereas Zn2+ is more challenging. When $Zn|Zn^{2+}$ is connected to $Cu|Cu^{2+}$ half-cells, the spontaneous process is the flow of e^- from $Zn|Zn^{2+}$ to $Cu|Cu^{2+}$ half-cells.

Half-reactions

A half-reaction is the oxidation or reduction reaction component of a redox reaction. A half-reaction is derived from the change in oxidation states of individual substances in the redox reaction.

Balancing redox reactions requires splitting the equation into the two half-reactions of reduction and oxidation.

In general, the half-reactions with the most positive (or the least negative) reduction potential occur before others.

Any duplication (copies, uploads, PDFs) is illegal.

Kinetic factor drives the reaction; if the difference in standard reduction potential between two half-cell reactions is small, the concentration may determine the outcome of electrolysis.

For example, the following reactions are possible for the electrolysis of 1 M NaCl (*aq*):

At anode:	$2\ Cl^- (aq) \rightarrow Cl_2 (g) + 2\ e^-$	$E° = -1.36$ V
	$2\ H_2O \rightarrow O_2 (g) + 4\ H^+ (aq) + 4\ e^-$	$E° = -1.23$ V
At cathode:	$2\ H_2O + 2\ e^- \rightarrow H_2 (g) + 2\ OH^- (aq)$	$E° = -0.83$ V
	$Na^+ (aq) + e^- \rightarrow Na (s)$	$E° = -2.71$ V

In the electrolysis of aqueous sodium chloride, Cl_2 forms at the anode instead of O_2, although E° (−1.23 V) for the formation of O_2 from H_2O is less negative than for the formation of Cl_2 from Cl^- (−1.36 V).

Formation of O_2 involves a higher activation energy (overvoltage); thus, it is kinetically less favorable.

H_2 is formed at the cathode since it requires less voltage than the reduction of $Na^+(aq)$.

For example, suppose a solution containing metal cations that require a lower potential (or one with a positive reduction potential) is electrolyzed.

Metal deposits on the cathode, and no hydrogen is produced. Electrolysis of an aqueous copper(II) chloride solution yields Cl_2 gas at the anode and deposits of copper metal at the cathode.

If the anode is copper, it is oxidized due to the lower voltage requirements compared to water.

Anode reaction:	$Cu (s) \rightarrow Cu^{2+} (aq) + 2\ e^-$	$E° = -0.34$ V
Cathode reaction:	$Cu^{2+} (aq) + 2\ e^- \rightarrow Cu (s)$	$E° = +0.34$ V

 Copyright © Sterling Education.

Electrical Potential Difference

Cell potential

Electrons flow from one electrode to another in a galvanic cell because there is an electrical potential difference between the two half-cells, called the *cell potential.*

Magnitude of cell potential depends on the two half-cells' nature, the concentration of electrolytes in each half-cell, and temperature.

Cell potential for galvanic (voltaic) cells is *positive* because the voltaic cell generates a potential difference.

Cell potential for electrolytic cells is *negative* because it requires an input of potential energy.

Standard cell potential, $E°_{cell}$, is measured under standard conditions (1 atm pressure for gas, 1 M of electrolytes, and 25 °C). The parameters for a standard cell require the species in solution to have a concentration of 1 M, the partial pressure of gases to be 1 atm, and the temperature to be 25 °C (298 K).

Reduction potential = potential of the reduction half-reaction

Oxidation potential = potential of the oxidation half-reaction

Cell potential = reduction potential + oxidation potential

In the zinc-copper cell, electrons flow from $Zn|Zn^{2+}$ half-cell to $Cu|Cu^{2+}$ half-cell. The $Zn|Z^{2+}$ half-cell has a higher electrical potential than $Cu|Cu^{2+}$.

Using these conditions makes it possible to describe a standard electrode potential (E°) in volts and calculate the cell's electrical potential. The superscript ° denotes standard conditions.

Cell potential and Gibbs free energy

Cell potential measures the potential difference between half-cells. It indicates if the cell reaction is spontaneous or nonspontaneous. Spontaneity for a process is often associated with its change in Gibbs free energy (ΔG).

Mathematical relationship between cell potential and ΔG is:

$\Delta G° = -nFE°$ under standard conditions

$\Delta G = -nFE$ under nonstandard conditions

where ΔG = change in Gibbs free energy in Joules, n = the moles of electrons transferred, the Faraday's constant F = 96,485 C/mol e^-, E = cell potential

Any duplication (copies, uploads, PDFs) is illegal.

Maximum work produced

Faraday's constant (C/mol e^-) and volts (V) for cell potential gas ΔG units of coulomb-volts.

Coulomb-volts are equivalent to *Joules*.

A potential difference of 1 V is equivalent to 1 Joule of *work done per coulomb of charge* that flows between two points in the circuit. (1 V = 1 J/C or 1 J = 1 C·V)

Maximum work produced = charge × maximum potential

$$w_{max} = -qE_{max} = \Delta G$$

Electrical charge

$$q = nF$$

$$\Delta G = -nFE_{cell}$$

or

$$\Delta G° = -nFE°_{cell}$$

where n = mole of electrons transferred or that flow through circuit, Faraday's constant F = 96,485 C/mole e^-

For example:

$$Zn\,(s) + Cu^{2+}(aq) \rightarrow Zn^{2+}(aq) + Cu\,(s) \qquad E°_{cell} = 1.10\text{ V}$$

$$\Delta G° = -2\text{ mol }e^- \times (96{,}485\text{ C / mol }e^-) \times 1.10\text{ V}$$

$$\Delta G° = -2.12 \times 10^5\text{ J}$$

2.12×10^2 kJ is the maximum work per mole of Zn reacted by Cu^{2+}.

Net cell potential

Net cell potential is the sum of the two half-cell potentials.

For the Zn-copper cell, the two half-cell potentials:

Anode half-cell reaction:

$$Zn\,(s) \rightarrow Zn^{2+}(aq) + 2\,e^- \qquad E°_{Zn \rightarrow Zn2+} = 0.76\text{ V}$$

Cathode half-cell reaction:

$$Cu^{2+}(aq) + 2\,e^- \rightarrow Cu\,(s) \qquad E°_{Cu2+ \rightarrow Cu} = 0.34\text{ V}$$

Copyright © Sterling Education.

Overall cell reaction:

$$Zn\ (s) + Cu^{2+}\ (aq) \rightarrow Zn^{2+}\ (aq) + Cu\ (s)$$

$$E°_{cell} = E°_{Zn \rightarrow Zn2+} + E°_{Cu2+ \rightarrow Cu}$$

$$E°_{cell} = 0.76\ V + 0.34\ V$$

$$E°_{cell} = 1.10\ V$$

For Cr|Cr^{3+} and Ag|Ag^+ half-cells, cell notation and potential:

Cr|Cr^{3+} (*aq*,1 M)||Ag^+ (*aq*,1 M)|Ag

(anode) (cathode)

Anode half-cell reaction:

$Cr\ (s) \rightarrow Cr^{3+}\ (aq) + 3\ e^-$ $\quad E°_{Cr \rightarrow Cr3+} = 0.73\ V$

Cathode half-cell reaction:

$Ag^+\ (aq) + e^- \rightarrow Ag\ (s)$ $\quad E°_{Ag+ \rightarrow Ag} = 0.80\ V$

Overall cell reaction:

$$Cr\ (s) + 3\ Ag^+\ (aq) \rightarrow Cr^{3+}\ (aq) + 3\ Ag\ (s)$$

$$E°_{cell} = E°_{Cr \rightarrow Cr3+} + E°_{Ag+ \rightarrow Ag}$$

$$E°_{cell} = 0.73\ V + 0.80\ V$$

$$E°_{cell} = 1.53\ V$$

Standard electrode potentials are typically listed as reduction potentials relative to a standard hydrogen electrode (SHE) with a potential of 0 volts.

For example, use the following metal ion/metal reaction potentials to calculate the standard cell potential:

$$Co\ (s) + Cu^{2+}\ (aq) \rightarrow Co^{2+}\ (aq) + Cu\ (s)$$

Cu^{2+} (*aq*)|Cu (*s*)	Ag^+ (*aq*)|Ag (*s*)	Co^{2+} (*aq*)|Co (*s*)	Zn^{2+} (*aq*)|Zn (*s*)
+0.34 V	+0.80 V	–0.28 V	–0.76 V

The cell reaction is:

$$Co\ (s) + Cu^{2+}\ (aq) \rightarrow Co^{2+}\ (aq) + Cu\ (s)$$

Separate it into half reactions:

$Cu^{2+}\ (aq) \rightarrow Cu\ (s)$ $\quad Co\ (s) \rightarrow Co^{2+}\ (aq)$

Any duplication (copies, uploads, PDFs) is illegal.

The potentials provided are written in this format:

Cu^{2+} (*aq*) | Cu (*s*)

+0.34 V

It means that for the reduction reaction:

Cu^{2+} (*aq*) → Cu (*s*), the potential is +0.34 V

The reverse reaction or oxidation reaction is:

Cu (*s*) → Cu^{2+} (*aq*) has opposing potential value: –0.34 V

Obtain potential values for each half-reaction.

Reverse sign for potential values of oxidation reactions:

Cu^{2+} (*aq*) → Cu (*s*) = +0.34 V (reduction)

Co (*s*) → Co^{2+} (*aq*) = +0.28 V (oxidation)

Determine the standard cell potential:

Standard cell potential = sum of half-reaction potentials

Standard cell potential = 0.34 V + 0.28 V

Standard cell potential = 0.62 V

Nernst equation

Nernst Equation is used when the temperature is standard (298 K) and employs the natural logarithm (ln) or the common logarithm, base 10 (log).

Nernst equation is used for a cell not under standard conditions:

$$E_{cell} = E_{cell}° - (RT / nF) \ln Q$$

where E_{cell} = nonstandard cell potential in volts; $E°_{cell}$ = standard cell potential in volts; R = constant, 8.314 J/mol·K, T = temperature in K; n = moles of electrons transferred; F = Faraday's constant = 96,485 C/mol e^- (C represents coulombs) and Q = the reaction quotient expression.

$$E_{cell} = E°_{cell} - [(0.0257\ V) / n] \ln Q$$

or

$$E_{cell} = E°_{cell} - [(0.0592\ V) / n] \log Q$$

Copyright © Sterling Education.

Modifying one of the previous cells, as follows:

Zn (s) | Zn^{2+} (*aq*), 3.00M; SO_4^{2-} (3.00M) || Cu^{2+} (*aq*), 0.0015M; SO_4^{2-} (0.0015M) | Cu (s)

$E°_{cell} = +1.10V$

For standard temperature (298 K), the Nernst equation expression:

$E_{cell} = E°_{cell} - [(0.0257 \text{ V}) / n] \ln [Zn^{2+}] / [Cu^{2+}]$

$E_{cell} = +1.10 \text{ V} - (0.0257 \text{ V} / 2 \text{ mol } e^-) \ln (3.00 \text{ M} / 0.0015 \text{ M})$

Evaluating the above expression gives:

$E_{cell} = +1.10 \text{ V} - 0.098 \text{ V}$

$E_{cell} = 1.00 \text{ V}$

Table of standard reduction potentials

Reduction potential measures the tendency of a chemical species to acquire electrons and become reduced (i.e., the gain of electrons is a reduction).

Reduction potential is measured in volts (V) or millivolts (mV).

Positive E°, the higher the tendency for the species to be reduced (i.e., a strong oxidizing agent).

Negative E°, the higher the tendency for the species to be oxidized (i.e., strong reducing agent).

Standard Reduction Potentials in Aqueous Solution at 25 °C *	
Acidic Solutions	**E° (V)**
F_2 (*g*) + 2 e^- → 2 F^- (*aq*)	+2.87
Co^{3+} (*aq*) + e^- → Co^{2+} (*aq*)	+1.82
Pb^{4+} (*aq*) + 2 e^- → Pb^{2+} (*aq*)	+1.8
H_2O_2 (*aq*) + 2 H^+ (*aq*) + 2 e^- → 2 H_2O	+1.77
NiO_2 (*s*) + 4 H^+ (*aq*) + 2 e^- → Ni^{2+} (*aq*) + 2 H_2O	+1.7
PbO_2 (*s*) + SO_4^{2-} (*aq*) + 4 H^+ (*aq*) + 2 e^- → $PbSO_4$ (*s*) + 2 H_2O	+1.69
Au^+ (*aq*) + e^- → Au (*s*)	+1.68
2 HClO (*aq*) + 2 H^+ (*aq*) + 2 e^- → Cl_2 (g) + 2 H_2O	+1.63
Ce^{4+} (*aq*) + e^- → Ce^{3+} (*aq*)	+1.61

Any duplication (copies, uploads, PDFs) is illegal.

$NaBiO_3$ (*s*) + 6 H^+ (*aq*) + 2 e^- → Bi^{3+} (*aq*) + Na^+ (*aq*) + 3 H_2O	+1.6
MnO_4^- (*aq*) + 8 H^+ (*aq*) + 5 e^- → Mn^{2+} (*aq*) + 4 H_2O	+1.51
Au^{3+} (*aq*) + 3 e^- → Au (*s*)	+1.5
ClO_3^- (*aq*) + 6 H^+ (*aq*) + 5 e^- → 1/2 Cl_2 (*g*) + 3 H_2O	+1.47
BrO_3^- + 6 H^+ (*aq*) + 6 e^- → Br^- (*aq*) + 3 H_2O	+1.44
Cl_2 (*g*) + 2 e^- → 2 Cl^- (*aq*)	+1.36
$Cr_2O_7^{2-}$ + 14 H^+ (*aq*) + 6 e^- → 2 Cr^{3+} (*aq*) + 7 H_2O	+1.33
$N_2H_5^+$ (*aq*) + 3 H^+ (*aq*) + 2 e^- → 2 NH_4^+ (*aq*)	+1.24
MnO_2 (*s*) + 4 H^+ (*aq*) + 2 e^- → Mn^{2+} (*aq*) + 2 H_2O	+1.23
O_2 (*g*) + 4 H^+ (*aq*) + 4 e^- → 2 H_2O	+1.23
Pt^{2+} (*aq*) + 2 e^- → Pt (*s*)	+1.2
IO_3^- (*aq*) + 6 H^+ (*aq*) + 5 e^- → ½ I_2 (*aq*) + 3 H_2O	+1.19
ClO_4^- (*aq*) + 2 H^+ (*aq*) + 2 e^- → ClO_3^- (*aq*) + H_2O	+1.19
Br_2 (*l*) + 2 e^- → 2 Br^- (*aq*)	+1.07
$AuCl_4^-$ + 3 e^- → Au (*s*) + 4 Cl^- (*aq*)	+1
Pd^{2+} (*aq*) + 2 e^- → Pd (*s*)	+0.99
NO_3^- (*aq*) + 4 H^+ (*aq*) + 3 e^- → NO (*g*) + 2 H_2O	+0.96
NO_3^- (*aq*) + 3 H^+ (*aq*) + 2 e^- → HNO_2 (*aq*) + H_2O	+0.94
2 Hg^{2+} (*aq*) + 2 e^- → Hg_2^{2+} (*aq*)	+0.92
Hg^{2+} (*aq*) + 2 e^- → Hg (*l*)	+0.86
Ag^+ (*aq*) + e^- → Ag (*s*)	+0.79
Hg_2^{2+} (*aq*) + 2 e^- → 2 Hg (*l*)	+0.79
Fe^{3+} (*aq*) + e^- → Fe^{2+} (*aq*)	+0.77
$SbCl_6^-$ (*aq*) + 2 e^- → $SbCl_4^-$ (*aq*) + 2 Cl^- (*aq*)	+0.75
$[PtCl_4]^{2-}$ (*aq*) + 2 e^- → Pt (*s*) + 4 Cl^- (*aq*)	+0.73

Copyright © Sterling Education.

$O_2\,(g) + 2\,H^+\,(aq) + 2\,e^- \rightarrow H_2O_2\,(aq)$	+0.68
$[PtCl_6]^{2-}\,(aq) + 2\,e^- \rightarrow [PtCl_4]^{2-}\,(aq) + 2\,Cl^-\,(aq)$	+0.68
$H_3AsO_4\,(aq) + 2\,H^+\,(aq) + 2\,e^- \rightarrow H_3AsO_3\,(aq) + H_2O$	+0.58
$I_2\,(s) + 2\,e^- \rightarrow 2\,I^-\,(aq)$	+0.54
$TeO_2\,(s) + 4\,H^+\,(aq) + 4\,e^- \rightarrow Te\,(s) + 2\,H_2O$	+0.53
$Cu^+\,(aq) + e^- \rightarrow Cu\,(s)$	+0.521
$[RhCl_6]^{3-}\,(aq) + 3\,e^- \rightarrow Rh\,(s) + 6\,Cl^-\,(aq)$	+0.44
$Cu^{2+}\,(aq) + 2\,e^- \rightarrow Cu\,(s)$	+0.34
$HgCl_2\,(s) + 2\,e^- \rightarrow 2\,Hg\,(l) + 2\,Cl^-\,(aq)$	+0.27
$AgCl\,(s) + e^- \rightarrow Ag\,(s) + Cl^-\,(aq)$	+0.22
$SO_4^{2-}\,(aq) + 4\,H^+\,(aq) + 2\,e^- \rightarrow SO_2\,(g) + 2\,H_2O$	+0.2
$SO_4^{2-}\,(aq) + 4\,H^+\,(aq) + 2\,e^- \rightarrow H_2SO_3\,(g) + H_2O$	+0.17
$Cu^{2+}\,(aq) + e^- \rightarrow Cu^+\,(aq)$	+0.15
$Sn^{4+}\,(aq) + 2\,e^- \rightarrow Sn^{2+}\,(aq)$	+0.15
$S\,(s) + 2\,H^+\,(aq) + 2\,e^- \rightarrow H_2S\,(aq)$	+0.14
$AgBr\,(s) + e^- \rightarrow Ag\,(s) + Br^-\,(aq)$	+0.07
$2\,H^+\,(aq) + 2\,e^- \rightarrow H_2\,(g)$ (reference electrode)	0.0
$N_2O\,(g) + 6\,H^+\,(aq) + H_2O + 4\,e^- \rightarrow 2\,NH_3OH^+\,(aq)$	−0.05
$Pb^{2+}\,(aq) + 2\,e^- \rightarrow Pb\,(s)$	−0.13
$Sn^{2+}\,(aq) + 2\,e^- \rightarrow Sn\,(s)$	−0.14
$AgI\,(s) + e^- \rightarrow Ag\,(s) + I^-\,(aq)$	−0.15
$Sn^{4+}\,(aq) + 2\,e^- \rightarrow Sn^{2+}\,(aq)$	+0.15
$S\,(s) + 2\,H^+\,(aq) + 2\,e^- \rightarrow H_2S\,(aq)$	+0.14
$AgBr\,(s) + e^- \rightarrow Ag\,(s) + Br^-\,(aq)$	+0.07
$2\,H^+\,(aq) + 2\,e^- \rightarrow H_2\,(g)$ (reference electrode)	0.0

Any duplication (copies, uploads, PDFs) is illegal.

N_2O (*g*) + 6 H^+ (*aq*) + H_2O + 4 e^- → 2 NH_3OH^+ (*aq*)	−0.05
Pb^{2+} (*aq*) + 2 e^- → Pb (*s*)	−0.13
Sn^{2+} (*aq*) + 2 e^- → Sn (*s*)	−0.14
AgI (*s*) + e^- → Ag (*s*) + I^- (*aq*)	−0.15
$[SnF_6]^{2-}$ (*aq*) + 4 e^- → Sn (*s*) + 6 F^- (*aq*)	−0.25
Ni^{2+} (*aq*) + 2 e^- → Ni (*s*)	−0.25
Co^{2+} (*aq*) + 2 e^- → Co (*s*)	−0.28
Tl^+ (*aq*) + e^- → Tl (*s*)	−0.34
$PbSO_4$ (*s*) + 2 e^- → Pb (*s*) + SO_4^{2-} (*aq*)	−0.36
Se (*s*) + 2 H^+ (*aq*) + 2 e^- → H_2Se (*aq*)	−0.4
Cd^{2+} (*aq*) + 2 e^- → Cd (*s*)	−0.40
Cr^{3+} (*aq*) + e^- → Cr^{2+} (*aq*)	−0.41
Fe^{2+} (*aq*) + 2 e^- → Fe (*s*)	−0.44
2 CO_2 (*g*) + 2 H^+ (*aq*) + 2 e^- → $(COOH)_2$ (*aq*)	−0.49
Ga^{3+} (*aq*) + 3 e^- → Ga (*s*)	−0.53
HgS (*s*) + 2 H^+ (*aq*) + 2 e^- → Hg (*l*) + H_2S (*g*)	−0.72
Cr^{3+} (*aq*) + 3 e^- → Cr (*s*)	−0.74
Zn^{2+} (*aq*) + 2 e^- → Zn (*s*)	−0.76
2 H_2O (*l*) + 2 e^- → H_2 (*g*) + 2 OH^- (*aq*)	−0.83
Cr^{2+} (*aq*) + 2 e^- → Cr (*s*)	−0.91
Mn^{2+} (*aq*) + 2 e^- → Mn (*s*)	−1.18
V^{2+} (*aq*) + 2 e^- → V (*s*)	−1.18
Zr^{4+} (*aq*) + 4 e^- → Zr (*s*)	−1.53
Al^{3+} (*aq*) + 3 e^- → Al (*s*)	−1.66
H_2 (*g*) + 2 e^- → 2 H^- (*aq*)	−2.25

Copyright © Sterling Education.

Mg^{2+} (*aq*) + 2 e^- → Mg (*s*)	−2.37
Na^+ (*aq*) + e^- → Na (*s*)	−2.71
Ca^{2+} (*aq*) + 2 e^- → Ca (*s*)	−2.87
Sr^{2+} (*aq*) + 2 e^- → Sr (*s*)	−2.89
Ba^{2+} (*aq*) + 2 e^- → Ba (*s*)	−2.9
Rb^+ (*aq*) + e^- → Rb (*s*)	−2.93
K^+ (*aq*) + e^- → K (*s*)	−2.93
Li^+ (*aq*) + e^- → Li (*s*)	−3.05
Basic Solutions	**E° (V)**
ClO^- (*aq*) + H_2O + 2 e^- → Cl^- (*aq*) + 2 OH^- (*aq*)	+0.89
OOH^- (*aq*) + H_2O + 2 e^- → 3 OH^- (*aq*)	+0.88
2 NH_2OH (*aq*) + 2 e^- → N_2H_4 (*aq*) + 2 OH^- (*aq*)	+0.74
ClO_3^- (*aq*) + 3 H_2O + 6 e^- → Cl^- (*aq*) + 6 OH^- (*aq*)	+0.62
MnO_4^- (*aq*) + 2 H_2O + 3 e^- → MnO_2 (*s*) + 4 OH^- (*aq*)	+0.59
MnO_4^- (*aq*) + e^- → MnO_4^{2-} (*aq*)	+0.56
NiO_2 (*s*) + 2 H_2O + 2 e^- → $Ni(OH)_2$ (s) + 2 OH^- (*aq*)	+0.49
Ag_2CrO_4 (*s*) + 2 e^- → 2 Ag (s) + CrO_4^{2-} (*aq*)	+0.45
O_2 (*g*) + 2 H_2O + 4 e^- → 4 OH^- (*aq*)	+0.4
ClO_4^- (*aq*) + H_2O + 2 e^- → ClO_3^- (*aq*) + 2 OH^- (*aq*)	+0.36
Ag_2O (*s*) + H_2O + 2 e^- → 2 Ag (*s*) + 2 OH^- (*aq*)	+0.34
2 NO_2^- (*aq*) + 3 H_2O + 4 e^- → N_2O (*g*) + 6 OH^- (*aq*)	+0.15
N_2H_4 (*aq*) + 2 H_2O + 2 e^- → 2 NH_3 (*aq*) + 2 OH^- (*aq*)	+0.1
$[Co(NH_3)_6]^{3+}$ (*aq*) + e^- → $[Co(NH_3)_6]^{2+}$ (*aq*)	+0.1
HgO (*s*) + H_2O + 2 e^- → Hg (*l*) + 2 OH^- (*aq*)	+0.09
O_2 (*g*) + H_2O + 2 e^- → OOH^- (*aq*) + OH^- (*aq*)	+0.08

Any duplication (copies, uploads, PDFs) is illegal.

$NO_3^-\,(aq) + H_2O + 2\,e^- \rightarrow NO_2^-\,(aq) + 2\,OH^-\,(aq)$	+0.01
$MnO_2\,(s) + 2\,H_2O + 2\,e^- \rightarrow Mn(OH)_2\,(s) + 2\,OH^-\,(aq)$	−0.05
$CrO_4^{2-}\,(aq) + 4\,H_2O + 3\,e^- \rightarrow Cr(OH)_3\,(s) + 5\,OH^-\,(aq)$	−0.12
$Cu(OH)_2\,(s) + 2\,e^- \rightarrow Cu\,(s) + 2\,OH^-\,(aq)$	−0.36
$Fe(OH)_3\,(s) + e^- \rightarrow Fe(OH)_2\,(s) + OH^-\,(aq)$	−0.56
$2\,H_2O + 2\,e^- \rightarrow H_2\,(g) + 2\,OH^-\,(aq)$	−0.83
$2\,NO_3^-\,(aq) + 2\,H_2O + 2\,e^- \rightarrow N_2O_4\,(g) + 4\,OH^-\,(aq)$	−0.85
$Fe(OH)_2\,(s) + 2\,e^- \rightarrow Fe\,(s) + 2\,OH^-\,(aq)$	−0.88
$SO_4^{2-}\,(aq) + H_2O + 2\,e^- \rightarrow SO_3^{2-}\,(aq) + 2\,OH^-\,(aq)$	−0.93
$N_2\,(g) + 4\,H_2O + 4\,e^- \rightarrow N_2H_4\,(aq) + 4\,OH^-\,(aq)$	−1.15
$[Zn(OH)_4]^{2-}\,(aq) + 2\,e^- \rightarrow Zn\,(s) + 4\,OH^-\,(aq)$	−1.22
$Zn(OH)_2\,(s) + 2\,e^- \rightarrow Zn\,(s) + 2\,OH^-\,(aq)$	−1.25
$[Zn(CN)_4]^{2-}\,(aq) + 2\,e^- \rightarrow Zn\,(s) + 4\,CN^-\,(aq)$	−1.26
$Cr(OH)_3\,(s) + 3\,e^- \rightarrow Cr\,(s) + 3\,OH^-\,(aq)$	−1.3
$SiO_3^{2-}\,(aq) + 3\,H_2O + 4\,e^- \rightarrow Si\,(s) + 6\,OH^-\,(aq)$	−1.7

**table is for reference only, and values are approximate*

Copyright © Sterling Education.

Electron Flow

Direction of electron flow

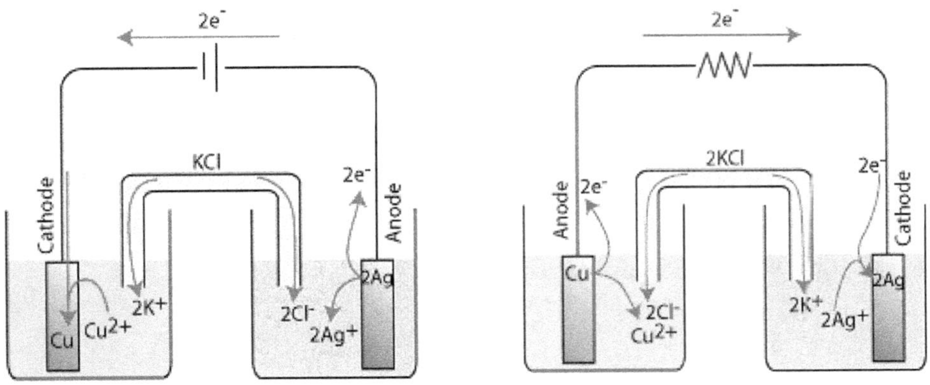

Electrolytic charge flow *Galvanic (voltaic) charge flow*

Electrons flow from anode to cathode.

Mnemonic: *A to C in alphabetical order*.

Alternatively, AC power, where the A comes first, stands for *Anode*.

Oxidation (at the anode) produces electrons (and cations) and releases electrons toward cathode.

Cathode receives those electrons and uses them for reduction.

Species with the highest oxidation potential (lowest reduction potential) is the anode, and the species with the highest reduction potential is the cathode.

In the diagram above, the galvanic (voltaic) cell exhibits a natural flow because Cu (with a higher oxidation potential/lower reduction potential) serves as the anode, and Ag (with a higher reduction potential) acts as the cathode.

Electrolytic cells show the opposite; a battery drives the reaction to force Cu to be the cathode and Ag to be the anode. Electrons flow in wires and electrodes, while ions flow in an electrolyte solution, thus creating a complete circuit.

Oxidation-reduction reaction continues until electrolyte concentrations in half-cells are equal.

Anode half-cell:

$Cu\ (s) \rightarrow Cu^{2+}\ (aq) + 2\ e^-$ (in 0.0010 M Cu^{2+})

Cathode half-cell:

$Cu^{2+}\ (aq) + 2\ e^- \rightarrow Cu\ (s)$ (in 0.50 M Cu^{2+})

Any duplication (copies, uploads, PDFs) is illegal.

Electromotive force and voltage

Electromotive force (emf), also known as cell potential (Ecell), is the driving force that enables electrons to flow from one electrode to another, measured in volts (V).

A volt is one Joule/coulomb, where the coulomb is the unit of charge.

Voltage is the driving force for reactions, often referred to as the electromotive force (EMF).

The higher the voltage, the greater the driving force.

Voltmeter must have its negative terminal connected to the anode and positive terminal to the cathode for the correct polarity (± sign).

Voltage is positive for a spontaneous reaction.

 Copyright © Sterling Education.

Batteries

Batteries as galvanic cells

Batteries are galvanic cells connected in series, where the battery potential equals the sum of the cells' potential.

Three types of batteries:

Primary batteries are not rechargeable.

Secondary batteries are rechargeable.

Fuel cells require a sufficient supply of fuel to generate energy.

Primary batteries include common (acidic), dry cells, alkaline batteries, and mercury batteries.

A new dry cell battery has a potential of approximately 1.5 V, regardless of its size; however, the amount of energy a battery can deliver depends on its size. D batteries deliver more current (greater amperes) than AAA batteries.

Standard dry batteries use aqueous NH_4Cl paste as an electrolyte and are acidic batteries due to the ionization:

$$NH_4^+\,(aq) \rightarrow NH_3\,(aq) + H^+\,(aq)$$

Acidic dry batteries consist of a Zinc casing as an anode, the reducing agent; a graphite rod as an (inert) cathode; aqueous NH_4Cl paste as an electrolyte, and MnO_2 powder as oxidizing agent.

$$Zn|Zn^{2+},NH_4^+,NH_3\,(aq)||Mn_2O_3,MnO_2|C\,(s)$$

Anode reaction:

$$Zn\,(s) \rightarrow Zn^{2+}\,(aq) + 2\,e^-$$

Cathodic reaction:

$$2\,MnO_2\,(s) + 2\,NH_4^+\,(aq) + 2\,e^- \rightarrow Mn_2O_3\,(s) + 2\,NH_3\,(aq) + H_2O\,(l)$$

Net reaction:

$$Zn\,(s) + 2\,MnO_2\,(s) + 2\,NH_4^+\,(aq) \rightarrow Zn^{2+}\,(aq) + Mn_2O_3\,(s) + 2\,NH_3\,(aq) + H_2O\,(l)$$

The reverse reaction is prevented by forming $[Zn(NH_3)_4]^{2+}$ ions.

Any duplication (copies, uploads, PDFs) is illegal.

Alkaline batteries utilize zinc (reducing agent) and MnO_2 (oxidizing agent), along with an aqueous KOH paste (not NH_4Cl) as the electrolyte.

Reaction at anode: $Zn\ (s) + 2\ OH^-\ (aq) \rightarrow ZnO\ (s) + H_2O\ (l) + 2\ e^-$

Reaction a cathode: $2\ MnO_2\ (s) + H_2O\ (l) + 2\ e^- \rightarrow Mn_2O_3\ (s) + 2\ OH^-\ (aq)$

Net reaction: $Zn\ (s) + 2\ MnO_2\ (s) \rightarrow ZnO\ (s) + Mn_2O_3\ (s)$

Since the reactants involved are solid, alkaline batteries deliver a constant voltage until the limiting reactant is exhausted. They last longer because zinc metal corrodes more slowly under basic conditions.

Mercury batteries are used in calculators and watches.

The following reactions occur at the anode and cathode sections of the cell:

Anode reaction: $Zn\ (s) + 2\ OH^-\ (aq) \rightarrow ZnO\ (s) + H_2O + 2\ e^-$

Cathode reaction: $HgO\ (s) + H_2O + 2\ e^- \rightarrow Hg\ (l) + 2\ OH^-\ (aq)$

Net cell reaction: $Zn\ (s) + HgO\ (s) \rightarrow ZnO\ (s) + Hg\ (l)$

Lead-storage batteries

Lead-storage batteries are used in automobiles and contain sulfuric acid as an electrolyte. Each battery cell contains several lead alloy grids with alternating grids packed with lead metal and lead(IV) oxide, PbO_2. The standard 12-V car battery contains six cells connected in series.

Each set of grids, which serves as electrodes, is connected in parallel, allowing the cell to deliver more current. The amount of current delivered depends on the electrode's surface area. Each cell in a lead storage battery has a potential of about 2.01 V.

Spontaneous reactions occurring in a lead storage battery:

Anode reaction:

$Pb\ (s) + HSO_4^-\ (aq) \rightarrow PbSO_4\ (s) + H^+\ (aq) + 2\ e^-$

Cathode reaction:

$PbO_2\ (s) + 3\ H^+\ (aq) + HSO_4^-\ (aq) + 2\ e^- \rightarrow PbSO_4\ (s) + 2\ H_2O\ (m)$

Net reaction:

$Pb\ (s) + PbO_2\ (s) + 2\ H^+\ (aq) + 2\ HSO_4^-\ (aq) \rightarrow 2\ PbSO_4\ (s) + 2\ H_2O\ (l) \rightarrow$ discharging

---

 Copyright © Sterling Education.

The greatest advantage of lead-storage batteries (i.e., secondary batteries) is being rechargeable. However, while the car is driven, the battery obtains energy from the motor through the alternator. Starting the engine causes the discharge reaction to occur.

The following recharging reaction occurs:

$$2\ PbSO_4\ (s) + 2\ H_2O\ (l) \rightarrow Pb\ (s) + PbO_2\ (s) + 2\ H^+\ (aq) + 2\ HSO_4^-\ (aq)$$

Lead-storage batteries have a longer lifetime than other batteries and deliver a large amount of current and electrical energy within a short time.

Major disadvantages of lead-storage batteries:

1) Heavy, bulky, and relatively non-portable

2) Lead is a toxic metal, and disposal is an environmental problem

3) Battery must be kept upright because H_2SO_4 is corrosive

Nickel-cadmium batteries

Nickel-cadmium batteries are rechargeable batteries used in small appliances such as cordless phones. They contain cadmium as the anode and hydrated nickel oxide as the cathode.

Electrolyte is an aqueous KOH paste.

Anode reaction:

$$Cd\ (s) + 2\ OH^-\ (aq) \rightarrow Cd(OH)_2\ (s) + 2\ e^-$$

Cathode reaction:

$$2\ NiO(OH)\ (s) + 2\ H_2O\ (l) + 2\ e^- \rightarrow 2\ Ni(OH)_2\ (s) + 2\ OH^-\ (aq)$$

Net cell reaction:

$$Cd\ (s) + 2\ NiO(OH)\ (s) + 2\ H_2O\ (l) \rightarrow Cd(OH)_2\ (aq) + 2\ Ni(OH)_2\ (aq)$$

Any duplication (copies, uploads, PDFs) is illegal.

Notes for active learning

 Copyright © Sterling Education.

Notes for active learning

Notes for active learning

Copyright © Sterling Education.

Practice Questions
&
Detailed Explanations

Copyright © Sterling Education.

Practice Questions

Practice Set 1: Questions 1–20

1. Which is NOT true regarding the redox reaction occurring in a spontaneous electrochemical cell?

$$Cl_2\,(g) + 2\,Br^-\,(aq) \rightarrow Br_2\,(l) + 2\,Cl^-\,(aq)$$

A. Anions flow towards the anode
B. Electrons flow from the anode to the cathode
C. Cl_2 is reduced at the cathode
D. Br^- is oxidized at the cathode
E. Br^- is oxidized at the anode

2. What is the relationship between an element's ionization energy and its ability to function as an oxidizing agent? As ionization energy increases:

A. the ability of an element to function as an oxidizing agent remains the same
B. the ability of an element to function as an oxidizing agent decreases
C. the ability of an element to function as an oxidizing agent increases
D. the ability of an element to function as a reducing agent remains the same
E. the ability of an element to function as a reducing agent increases

3. How many electrons are needed to balance the following half-reaction $H_2S \rightarrow S_8$ in an acidic solution?

A. 14 electrons to the left side
B. 6 electrons to the right side
C. 12 electrons to the left side
D. 8 electrons to the right side
E. 16 electrons to the right side

4. Which is NOT true regarding the following redox reaction occurring in an electrolytic cell?

$$3\,C + 2\,Co_2O_3 \xrightarrow{\text{Electricity}} 4\,Co + 3\,CO_2$$

A. $CO_2\,(g)$ is produced at the anode
B. Co metal is produced at the anode
C. Oxidation half-reaction: $C + 2\,O^{2-} \rightarrow CO_2 + 4\,e^-$
D. Reduction half-reaction: $Co^{3+} + 3\,e^- \rightarrow Co$
E. Co metal is produced at the cathode

Any duplication (copies, uploads, PDFs) is illegal.

5. The anode in a galvanic cell attracts:

A. cations
B. neutral particles
C. anions
D. both anions and neutral particles
E. both cations and neutral particles

6. The electrode with the standard reduction potential of 0 V is assigned as the standard reference electrode and uses the half-reaction:

A. $2\ NH_4^+ (aq) + 2\ e^- \leftrightarrow H_2 (g) + 2\ NH_3 (g)$
B. $Ag^+ (aq) + e^- \leftrightarrow Ag (s)$
C. $Cu^{2+} (aq) + 2\ e^- \leftrightarrow Cu (s)$
D. $Zn^{2+} (aq) + 2\ e^- \leftrightarrow Zn (s)$
E. $2\ H^+ (aq) + 2\ e^- \leftrightarrow H_2 (g)$

7. In the reaction for a discharging nickel–cadmium (NiCd) battery, which substance is being oxidized?

$$Cd (s) + NiO_2 (s) + 2\ H_2O (l) \rightarrow Cd(OH)_2 (s) + Ni(OH)_2 (s)$$

A. H_2O
B. $Cd(OH)_2$
C. Cd
D. NiO_2
E. $Ni(OH)_2$

8. What happens at the anode if rust forms when Fe is in contact with H_2O?

$$4\ Fe + 3\ O_2 \rightarrow 2\ Fe_2O_3$$

A. Fe is reduced
B. Oxygen is reduced
C. Oxygen is oxidized
D. Fe is oxidized
E. None of the above

9. Which is true regarding the redox reaction occurring in the spontaneous electrochemical cell for

$$Cl_2 (g) + 2\ Br^- (aq) \rightarrow Br_2 (l) + 2\ Cl^- (aq)?$$

A. Electrons flow from the cathode to the anode
B. Cl_2 is oxidized at the cathode
C. Br^- is reduced at the anode
D. Cations in the salt bridge flow from the Br_2 half-cell to the Cl_2 half-cell
E. Br^- is reduced at the anode, and Cl_2 is oxidized at the cathode

Copyright © Sterling Education.

10. What is the term for a chemical reaction that involves electron transfer between two reacting substances?

A. Reduction reaction
B. Electrochemical reaction
C. Oxidation reaction
D. Half-reaction
E. Redox reaction

11. Using the following metal ion/metal reaction potentials,

Cu^{2+} (*aq*)|Cu (*s*)	Ag^{+} (*aq*)|Ag (*s*)	Co^{2+} (*aq*)|Co (*s*)	Zn^{2+} (*aq*)|Zn (*s*)
+0.34 V	+0.80 V	–0.28 V	–0.76 V

calculate the standard cell potential for the cell whose reaction is:

$$\text{Co}\ (s) + \text{Cu}^{2+}\ (aq) \rightarrow \text{Co}^{2+}\ (aq) + \text{Cu}\ (s)$$

A. +0.62 V
B. –0.62 V
C. +0.48 V
D. –0.48 V
E. +0.68 V

12. How are photovoltaic cells different from many other forms of solar energy?

A. Light is reflected, and the coolness of the shade is used to provide a temperature differential
B. Light is passively converted into heat
C. Light is converted into heat and then into steam
D. Light is converted into heat and then into electricity
E. Light is converted directly to electricity

13. What is the term for an electrochemical cell with a single electrode where oxidation or reduction can occur?

A. Half-cell
B. Voltaic cell
C. Dry cell
D. Electrolytic cell
E. None of the above

Any duplication (copies, uploads, PDFs) is illegal.

14. Which method could electrolysis be used to raise the hull of a sunken ship?

A. Electrolysis could only be used to raise the hull if the ship is made of iron. If so, the electrolysis of the iron metal might produce sufficient gas to lift the ship
B. The gaseous products of the electrolysis of H_2O are collected with bags attached to the hull of the ship and the inflated bags raise the ship
C. The electrolysis of the H_2O beneath the hull of the ship boils H_2O and creates upward pressure to raise the ship
D. An electric current passed through the hull of the ship produces electrolysis and the gases trapped in compartments of the vessel would push it upwards
E. Electrolysis of the ship's hull decreases its mass, and the reduced weight causes the ship to rise

15. What are the products for the single-replacement reaction Zn (s) + $CuSO_4$ (aq) → ?

A. CuO and $ZnSO_4$
B. CuO and $ZnSO_3$
C. Cu and $ZnSO_4$
D. Cu and $ZnSO_3$
E. No reaction

16. Which of the following is a true statement about the electrochemical reaction for an electrochemical cell with a cell potential of +0.36 V?

A. The reaction favors the formation of reactants and would be considered a galvanic cell
B. The reaction favors the formation of reactants and would be considered an electrolytic cell
C. The reaction is at equilibrium and is a galvanic cell
D. The reaction favors the formation of products and would be considered an electrolytic cell
E. The reaction favors the formation of products and would be considered a galvanic cell

17. Which of the following is a unit of electrical energy?

A. Coulomb
B. Joule
C. Watt
D. Ampere
E. Volt

Copyright © Sterling Education.

18. Which observation describes the solution near the cathode when an aqueous solution of sodium chloride is electrolyzed and hydrogen gas is evolved at the cathode?

A. Colored
B. Acidic
C. Basic
D. Frothy
E. Viscous

19. How many grams of Ag is deposited in the cathode of an electrolytic cell if a current of 3.50 A is applied to a solution of $AgNO_3$ for 12 minutes? (Use the molecular mass of Ag = 107.86 g/mol and the conversion of 1 mole $e^- = 9.65 \times 10^4$ C)

A. 0.32 g
B. 0.86 g
C. 2.82 g
D. 3.86 g
E. 4.38 g

20. In an electrochemical cell, which of the following statements is FALSE?

A. The anode is the electrode where oxidation occurs
B. A salt bridge provides electrical contact between the half-cells
C. The cathode is the electrode where reduction occurs
D. A spontaneous electrochemical cell is called a galvanic cell
E. All of the above are true

Any duplication (copies, uploads, PDFs) is illegal.

Notes for active learning

 Copyright © Sterling Education.

Practice Set 2: Questions 21–40

21. What is the relationship between an element's ionization energy and its ability to function as an oxidizing and reducing agent? Elements with high ionization energy are:

A. strong oxidizing and weak reducing agents
B. weak oxidizing and weak reducing agents
C. strong oxidizing and strong reducing agents
D. weak oxidizing and strong reducing agents
E. weak oxidizing and neutral reducing agents

22. In a basic solution, how many electrons are needed to balance the following half-reaction?

$$C_8H_{10} \rightarrow C_8H_4O_4^{2-}$$

A. 8 electrons on the left side
B. 12 electrons on the left side
C. 4 electrons on the left side
D. 12 electrons on the right side
E. 8 electrons on the right side

23. Which statement is true regarding the following redox reaction occurring in an electrolytic cell?

$$3\ C\ (s) + 2\ Co_2O_3\ (l) \xrightarrow{\text{Electricity}} 4\ Co\ (l) + 3\ CO_2\ (g)$$

A. CO_2 gas is produced at the anode
B. Co^{3+} is produced at the cathode
C. Oxidation half-reaction: $Co^{3+} + 3\ e^- \rightarrow Co$
D. Reduction half-reaction: $C + 2\ O^{2-} \rightarrow CO_2 + e^-$
E. None of the above

Any duplication (copies, uploads, PDFs) is illegal.

24. Which is true regarding the following redox reaction occurring in a spontaneous electrochemical cell?

$$Sn\ (s) + Cu^{2+}\ (aq) \rightarrow Cu\ (s) + Sn^{2+}\ (aq)$$

A. Anions in the salt bridge flow from the Cu half-cell to the Sn half-cell
B. Electrons flow from the Cu electrode to the Sn electrode
C. Cu^{2+} is oxidized at the cathode
D. Sn is reduced at the anode
E. None of the above

25. In galvanic cells, reduction occurs at the:

I. salt bridge II. cathode III. anode

A. I only
B. II only
C. III only
D. I and II only
E. I, II and III

26. For a battery, what is undergoing a reduction in the following oxidation-reduction reaction?

$$Mn_2O_3 + ZnO \rightarrow 2\ MnO_2 + Zn$$

A. Mn_2O_3
B. MnO_2
C. ZnO
D. Zn
E. Mn_2O_3 and Zn

27. Given that the following redox reactions go essentially to completion, which of the metals listed below has the greatest tendency to undergo oxidation?

$$Ni\ (s) + Ag^{+}\ (aq) \rightarrow Ag\ (s) + Ni^{2+}\ (aq)$$

$$Al\ (s) + Cd^{2+}\ (aq) \rightarrow Cd\ (s) + Al^{3+}\ (aq)$$

$$Cd\ (s) + Ni^{2+}\ (aq) \rightarrow Ni\ (s) + Cd^{2+}\ (aq)$$

$$Ag\ (s) + H^{+}\ (aq) \rightarrow \text{no reaction}$$

A. (H) **B.** Cd **C.** Ni **D.** Ag **E.** Al

Copyright © Sterling Education.

28. What is the purpose of the salt bridge in a voltaic cell?

A. Allows for a balance of charge between the two chambers
B. Allows the Fe^{2+} and the Cu^{2+} to flow freely between the two chambers
C. Allows for the buildup of positively charged ions in one container and negatively charged ions in the other container
D. Prevents any further migration of electrons through the wire
E. None of the above

29. If ΔG° for a cell is positive, the E° is:

A. neutral
B. negative
C. positive
D. unable to be determined
E. greater than 1

30. In an electrochemical cell, which of the following statements is FALSE?

A. Oxidation occurs at the anode
B. Reduction occurs at the cathode
C. Electrons flow through the salt bridge to complete the cell
D. A spontaneous electrochemical cell is called a voltaic cell
E. All of the above are true

31. Which of the following statements about electrochemistry is NOT true?

A. The study of how protons are transferred from one chemical compound to another
B. The use of electrical current to produce an oxidation-reduction reaction
C. The use of a set of oxidation-reduction reactions to produce electrical current
D. The study of how electrical energy and chemical reactions are related
E. The study of how electrons are transferred from one chemical compound to another

32. In an operating photovoltaic cell, electrons move through the external circuit to the negatively charged p-type silicon wafer. How can the electrons move to the negatively charged silicon wafer if electrons are negatively charged?

A. The p-type silicon wafer is positively charged
B. The energy of the sunlight moves electrons in a nonspontaneous direction
C. Advancements in photovoltaic technology have solved this technological impediment
D. An electric current occurs because energy from the sun reverses the charge of electrons
E. The p-type silicon wafer is negatively charged

Any duplication (copies, uploads, PDFs) is illegal.

33. What is the term for the value assigned to an atom in a substance that indicates whether the atom is electron-poor or electron-rich compared to a free atom?

A. Reduction number
B. Oxidation number
C. Cathode number
D. Anode number
E. None of the above

34. What is the term for the relative ability of a substance to undergo reduction?

A. Oxidation potential
B. Reduction potential
C. Anode potential
D. Cathode potential
E. None of the above

35. What is the primary difference between a fuel cell and a battery?

A. Fuel cells oxidize to supply electricity, while batteries reduce to supply electricity
B. Batteries supply electricity, while fuel cells supply heat
C. Batteries can be recharged, while fuel cells cannot
D. Fuel cells do not run down because they can be refueled, while batteries run down and need to be recharged
E. Fuel cells do not use metals as oxidants and reductants, while batteries have a static reservoir of oxidants and reductants

36. In balancing the equation for a disproportionation reaction, using the oxidation number method, the substance undergoing disproportionation is:

A. initially written twice on the reactant side of the equation
B. initially written twice on the product side of the equation
C. initially written on both the reactant and product sides of the equation
D. assigned an oxidation number of zero
E. assigned an oxidation number of +1

37. What is the term for an electrochemical cell in which electrical energy is generated from a spontaneous redox reaction?

A. Voltaic cell
B. Photoelectric cell
C. Dry cell
D. Electrolytic cell
E. Wet cell

38. Which observation describes the solution near the anode when an aqueous solution of sodium sulfate is electrolyzed and gas is evolved at the anode?

Copyright © Sterling Education.

A. Colored
B. Acidic
C. Basic
D. Frothy
E. Viscous

39. Which of the statements listed below is true regarding the following redox reaction occurring in a nonspontaneous electrochemical cell?

$$Cd\ (s) + Zn(NO_3)_2\ (aq) + \text{electricity} \rightarrow Zn\ (s) + Cd(NO_3)_2\ (aq)$$

A. Oxidation half-reaction: $Zn^{2+} + 2\ e^- \rightarrow Zn$
B. Reduction half-reaction: $Cd \rightarrow Cd^{2+} + 2\ e^-$
C. Cd metal is produced at the anode
D. Zn metal is produced at the cathode
E. None of the above

40. Which statement is true for electrolysis?

A. A spontaneous redox reaction produces electricity
B. A nonspontaneous redox reaction is forced to occur by applying an electric current
C. Only pure, drinkable water is produced
D. There is a cell that reverses the flow of ions
E. None of the above

Notes for active learning

Copyright © Sterling Education.

Practice Set 3: Questions 41–60

41. Which relationship explains an element's electronegativity and its ability to act as an oxidizing and reducing agent?

A. Atoms with large electronegativity tend to act as strong oxidizing and strong reducing agents
B. Atoms with large electronegativity tend to act as weak oxidizing and strong reducing agents
C. Atoms with large electronegativity tend to act as strong oxidizing and weak reducing agents
D. Atoms with large electronegativity tend to act as weak oxidizing and weak reducing agents
E. None of the above

42. Which substance is undergoing reduction for a battery if the following two oxidation-reduction reactions take place?

Reaction I: $Zn + 2\ OH^- \rightarrow ZnO + H_2O + 2\ e^-$

Reaction II: $2\ MnO_2 + H_2O + 2\ e^- \rightarrow Mn_2O_3 + 2\ OH^-$

A. OH^- **B.** H_2O **C.** Zn **D.** ZnO **E.** MnO_2

43. What is the term for an electrochemical cell when a nonspontaneous redox reaction occurs by forcing electricity through the cell?

I. voltaic cell II. dry cell III. electrolytic cell

A. I only
B. II only
C. III only
D. I and II only
E. I and III only

44. Which statement is true regarding the redox reaction occurring in a galvanic cell?

$Cl_2\ (g) + 2\ Br^-\ (aq) \rightarrow Br_2\ (l) + 2\ Cl^-\ (aq)$?

A. Cations in the salt bridge flow from the Cl_2 half-cell to the Br_2 half-cell
B. Electrons flow from the anode to the cathode
C. Cl_2 is reduced at the anode
D. Br^- is oxidized at the cathode
E. None of the above

45. Which of the statements below is true regarding the following redox reaction occurring in a nonspontaneous electrolytic cell?

$$3\ C\ (s) + 4\ AlCl_3\ (l) \xrightarrow{\text{Electricity}} 4\ Al\ (l) + 3\ CCl_4\ (g)$$

A. Cl^- is produced at the anode
B. Al metal is produced at the cathode
C. Oxidation half-reaction is $Al^{3+} + 3\ e^- \rightarrow Al$
D. Reduction half-reaction is $C + 4\ Cl^- \rightarrow CCl_4 + 4\ e^-$
E. None of the above

46. Which of the following materials is most likely to undergo oxidation?

I. Cl^- II. Na III. Na^+

A. I only
B. II only
C. III only
D. I and II only
E. I, II and III

47. Which statement is NOT true regarding the redox reaction occurring in a spontaneous electrochemical cell?

$$Zn\ (s) + Cd^{2+}\ (aq) \rightarrow Cd\ (s) + Zn^{2+}\ (aq)$$

A. Anions in the salt bridge flow from the Zn half-cell to the Cd half-cell
B. Electrons flow from the Zn electrode to the Cd electrode
C. Cd^{2+} is reduced at the cathode
D. Zn is oxidized at the anode
E. Anions in the salt bridge flow from the Cd half-cell to the Zn half-cell

48. Which is capable of oxidizing Cu (*s*) to Cu^{2+} (*aq*) when added to Cu (*s*) in solution?

A. Al^{3+} (*aq*)
B. Ag^+ (*aq*)
C. Ni (*s*)
D. I^- (*aq*)
E. Zn^{2+} (*aq*)

Copyright © Sterling Education.

49. Why is the anode of a battery indicated with a negative (–) sign?

A. Electrons move to the anode to react with NH_4Cl in the battery
B. It indicates the electrode where the chemicals are reduced
C. Electrons are attracted to the negative electrode
D. The cathode is the source of negatively charged electrons
E. The electrode is the source of negatively charged electrons

50. What is the term for converting chemical energy to electrical energy from redox reactions?

A. Redox chemistry
B. Electrochemistry
C. Cell chemistry
D. Battery chemistry
E. None of the above

51. In electrolysis, E° tends to be:

A. zero
B. neutral
C. positive
D. greater than 1
E. negative

52. A major source of chlorine gas is the electrolysis of concentrated saltwater, NaCl (*aq*). What is the sign of the electrode where the chlorine gas is formed?

A. Neither, since chlorine gas is a neutral molecule and there is no electrode attraction
B. Negative, since the chlorine gas needs to deposit electrons to form chloride ions
C. Positive, since the chloride ions lose electrons to form chlorine molecules
D. Both, since the chloride ions from NaCl (*aq*) are attracted to the positive electrode to form chlorine molecules, while the produced chlorine gas molecules move to deposit electrons at the negative electrode
E. Positive, since the chloride ions gain electrons to form chlorine molecules

53. Which statement is correct for an electrolytic cell with two electrodes?

A. Oxidation occurs at the anode, which is negatively charged
B. Oxidation occurs at the anode, which is positively charged
C. Oxidation occurs at the cathode, which is positively charged
D. Oxidation occurs at the cathode, which is negatively charged
E. Oxidation occurs at the dynode, which is uncharged

Any duplication (copies, uploads, PDFs) is illegal.

54. Which of the statements listed below is true regarding the following redox reaction occurring in a nonspontaneous electrochemical cell?

$$Cd\ (s) + Zn(NO_3)_2\ (aq) + \text{electricity} \rightarrow Zn\ (s) + Cd(NO_3)_2\ (aq)$$

A. Oxidation half-reaction: $Zn^{2+} + 2\ e^- \rightarrow Zn$
B. Reduction half-reaction: $Cd \rightarrow Cd^{2+} + 2\ e^-$
C. Cd metal dissolves at the anode
D. Zn metal dissolves at the cathode
E. None of the above

55. Which battery system is based on half-reactions involving zinc metal and manganese dioxide?

A. Alkaline batteries
B. Fuel cells
C. Lead-acid storage batteries
D. Dry-cell batteries
E. None of the above

56. Which of the following equations is a disproportionation reaction?

A. $2\ H_2O \rightarrow 2\ H_2 + O_2$
B. $H_2SO_3 \rightarrow H_2O + SO_2$
C. $HNO_2 \rightarrow NO + HNO_3$
D. $Mg + H_2SO_4 \rightarrow MgSO_4 + H_2$
E. None of the above

57. How is electrolysis different from the chemical process inside a battery?

A. Pure compounds cannot be generated *via* electrolysis
B. Electrolysis only uses electrons from a cathode
C. Electrolysis does not use electrons
D. They are the same process in reverse
E. Chemical changes do not occur in electrolysis

58. Which fact about fuel cells is FALSE?

A. Fuel cell automobiles are powered by water and only emit hydrogen
B. Fuel cells are based on the tendency of some elements to gain electrons from another
C. Fuel cell automobiles are quiet
D. Fuel cell automobiles are environmentally friendly
E. All of the above

Copyright © Sterling Education.

59. Which process occurs when copper is refined using the electrolysis technique?

A. Impure copper goes into solution at the anode, and pure copper plates out on the cathode
B. Impure copper goes into solution at the cathode, and pure copper plates out on the anode
C. Pure copper goes into solution from the anode and forms a precipitate at the bottom of tank
D. Pure copper goes into solution from the cathode and forms a precipitate at bottom of tank
E. Pure copper on the bottom of the tank goes into solution and plates out on the cathode

60. Which of the following statements describes electrolysis?

A. A chemical reaction which results when electrical energy is passed through a metallic liquid
B. A chemical reaction which results when electrical energy is passed through a liquid electrolyte
C. The splitting of atomic nuclei by electrical energy
D. The splitting of atoms by electrical energy
E. The passage of electrical energy through a split-field armature

Notes for active learning

Copyright © Sterling Education.

Practice Set 4: Questions 61–80

61. In a battery, which of the following species in the two oxidation-reduction reactions is undergoing oxidation?

Reaction I: $Zn + 2\ OH^- \rightarrow ZnO + H_2O + 2\ e^-$

Reaction II: $2\ MnO_2 + H_2O + 2\ e^- \rightarrow Mn_2O_3 + 2\ OH^-$

A. H_2O **B.** MnO_2 **C.** ZnO **D.** Zn **E.** OH^-

62. What is the term for an electrochemical cell where the anode and cathode reactions do not occur in aqueous solutions?

A. Voltaic cell
B. Electrolytic cell
C. Dry cell
D. Galvanic cell
E. None of the above

63. How many electrons are needed to balance the charge for the following half-reaction in an acidic solution?

$C_2H_6O \rightarrow HC_2H_3O_2$

A. 6 electrons to the right side
B. 4 electrons to the right side
C. 3 electrons to the left side
D. 2 electrons to the left side
E. 8 electrons to the right side

64. Which of the statements is true regarding the following redox reaction occurring in an electrolytic cell?

$$3\ C\ (s) + 4\ AlCl_3\ (l) \xrightarrow{\text{Electricity}} 4\ Al\ (l) + 3\ CCl_4\ (g)$$

A. Al^{3+} is produced at the cathode
B. CCl_4 gas is produced at the anode
C. Reduction half-reaction: $C + 4\ Cl^- \rightarrow CCl_4 + 4\ e^-$
D. Oxidation half-reaction: $Al^{3+} + 3\ e^- \rightarrow Al$
E. None of the above

Any duplication (copies, uploads, PDFs) is illegal.

65. What is the term for an electrochemical cell that spontaneously produces electrical energy?

I. half-cell   II. electrolytic cell   III. dry cell

A. I and II only
B. II only
C. III only
D. II and III only
E. None of the above

66. The anode of a battery is indicated with a negative (–) sign because the anode is:

A. where electrons are adsorbed
B. positive
C. negative
D. where electrons are generated
E. determined by convention

67. Which is the strongest reducing agent for the following half-reaction potentials?

Sn^{4+} (*aq*) + 2 e^- → Sn^{2+} (*aq*)   E° = –0.13 V

Ag^+ (*aq*) + e^- → Ag (*s*)   E° = +0.81 V

Cr^{3+} (*aq*) + 3 e^- → Cr (*s*)   E° = –0.75 V

Fe^{2+} (*aq*) + 2 e^- → Fe (*s*)   E° = –0.43 V

A. Cr (*s*)
B. Fe^{2+} (*aq*)
C. Sn^{2+} (*aq*)
D. Ag (*s*)
E. Sn^{2+} (*aq*) and Ag (*s*)

68. Which is true for the redox reaction occurring in a spontaneous electrochemical cell?

Zn (*s*) + Cd^{2+} (*aq*) → Cd (*s*) + Zn^{2+} (*aq*)

A. Electrons flow from the Cd electrode to the Zn electrode
B. Anions in the salt bridge flow from the Cd half-cell to the Zn half-cell
C. Zn is reduced at the anode
D. Cd^{2+} is oxidized at the cathode
E. None of the above

Copyright © Sterling Education.

69. Based upon the reduction potential: $Zn^{2+} + 2\ e^- \rightarrow Zn\ (s)$; $E° = -0.76$ V, does a reaction take place when Zinc (*s*) is added to aqueous HCl, under standard conditions?

A. Yes, because the reduction potential for H^+ is negative
B. Yes, because the reduction potential for H^+ is zero
C. No, because the oxidation potential for Cl^- is positive
D. No, because the reduction potential for Cl^- is negative
E. Yes, because the reduction potential for H^+ is positive

70. If 1 amp of current passes a cathode for 10 minutes, how much Zn (*s*) forms in the following reaction? (Use the molecular mass of Zn = 65 g/mole)

$$Zn^{2+} + 2\ e^- \rightarrow Zn\ (s)$$

A. 0.10 g
B. 10.0 g
C. 0.65 g
D. 2.20 g
E. 0.20 g

71. In the oxidation–reduction reaction $Mg\ (s) + Cu^{2+}\ (aq) \rightarrow Mg^{2+}\ (aq) + Cu\ (s)$, which atom/ion is reduced and which atom/ion is oxidized?

A. The Cu^{2+} ion is reduced (gains electrons) to form Cu metal, while Mg metal is oxidized (loses electrons) to form Mg^{2+}
B. Since Mg is transformed from a solid to an aqueous solution and Cu is transformed from an aqueous solution to a solid, no oxidation–reduction reaction occurs
C. The Cu^{2+} ion is oxidized (i.e., gains electrons) from Cu metal, while Mg metal is reduced (i.e., loses electrons) to form Mg^{2+}
D. The Mg^{2+} ion is reduced (i.e., gains electrons) from Cu metal, while Cu^{2+} is oxidized (i.e., loses electrons) to the Mg^{2+}
E. None of the above

72. Electrolysis is an example of a(n):

A. acid-base reaction
B. exothermic reaction
C. physical change
D. chemical change
E. state function

73. What is the term for a process characterized by the loss of electrons?

A. Redox
B. Reduction
C. Electrochemistry
D. Oxidation
E. None of the above

Any duplication (copies, uploads, PDFs) is illegal.

74. Which of the following statements is true about a galvanic cell?

A. The standard reduction potential for the anode reaction is always positive.
B. The standard reduction potential for the anode reaction is always negative.
C. The standard reduction potential for the cathode reaction is always positive.
D. E° for the cell is always positive.
E. E° for the cell is always negative.

75. Which of the following statements is true?

A. Galvanic cells were invented by Thomas Edison
B. Galvanic cells generate electrical energy rather than consume it
C. Electrolysis cells generate alternating current when their terminals are reversed
D. Electrolysis was discovered by Lewis Latimer
E. The laws of electrolysis were discovered by Granville Woods

76. Which statement is correct for an electrolysis cell with two electrodes?

A. Reduction occurs at the anode, which is positively charged
B. Reduction occurs at the anode, which is negatively charged
C. Reduction occurs at the cathode, which is positively charged
D. Reduction occurs at the cathode, which is negatively charged
E. Reduction occurs at the dynode, which is uncharged

77. How long would it take to deposit 4.00 grams of Cu from a $CuSO_4$ solution if a current of 2.5 A is applied? (Use the molecular mass of Cu = 63.55 g/mol and the conversion of 1 mole $e^- = 9.65 \times 10^4$ C and 1 C = A·s)

A. 1.02 hours
B. 1.36 hours
C. 4.46 hours
D. 6.38 hours
E. 7.72 hours

78. Which battery system is completely rechargeable?

A. Alkaline batteries
B. Fuel cells
C. Lead-acid storage batteries
D. Dry-cell batteries
E. None of the above

 Copyright © Sterling Education.

79. In which type of cell does the following reaction occur when electrons are forced into a system by applying an external voltage?

$Fe^{2+} + 2e^{-} \rightarrow Fe\ (s)$ $E° = -0.44\ V$

A. Concentration cell
B. Battery
C. Electrochemical cell
D. Galvanic cell
E. Electrolytic cell

80. What is the term for a reaction that represents separate oxidation or reduction processes?

A. Reduction reaction
B. Redox reaction
C. Oxidation reaction
D. Half-reaction
E. None of the above

Any duplication (copies, uploads, PDFs) is illegal.

Notes or active learning

Copyright © Sterling Education.

Answer Key & Detailed Explanations

Answer Key

1: D	11: A	21: A	31: A	41: C	51: E	61: D	71: A
2: C	12: E	22: D	32: B	42: E	52: C	62: C	72: D
3: E	13: A	23: A	33: B	43: C	53: B	63: B	73: D
4: B	14: B	24: A	34: B	44: B	54: C	64: B	74: D
5: C	15: C	25: B	35: D	45: B	55: D	65: C	75: B
6: E	16: E	26: C	36: A	46: B	56: C	66: D	76: D
7: C	17: B	27: E	37: A	47: A	57: D	67: A	77: B
8: D	18: C	28: A	38: B	48: B	58: A	68: B	78: C
9: D	19: C	29: B	39: D	49: E	59: A	69: B	79: E
10: E	20: E	30: C	40: B	50: B	60: B	70: E	80: D

Practice Set 1: Questions 1–20

1. D is correct.

In electrochemical (i.e., galvanic) cells, oxidation occurs at the anode, and reduction occurs at the cathode.

Br^- is oxidized at the anode, not the cathode.

2. C is correct.

As ionization energy increases, it is more difficult for electrons to be released by a substance. The release of electrons would increase the charge of the substance, which is the definition of oxidation.

Substances with higher ionization energy are more likely to be reduced.

If a substance is reduced in a redox reaction, it is considered an oxidizing agent because it facilitates the oxidation of the other reactant.

Therefore, with the increase of ionization energy, a substance will be more likely to be reduced or be a stronger oxidizing agent.

Any duplication (copies, uploads, PDFs) is illegal.

3. E is correct.

Half-reaction:

$$H_2S \rightarrow S_8$$

Balancing half-reaction in acidic conditions:

Step 1: Balance all atoms except for H and O

$$8\ H_2S \rightarrow S_8$$

Step 2: To balance oxygen, add H_2O to the side with fewer oxygen atoms

There is no oxygen at all, so skip this step.

Step 3: To balance hydrogen, add H^+:

$$8\ H_2S \rightarrow S_8 + 16\ H^+$$

Balance charges by adding electrons to the side with a higher/more positive total charge.

Total charge on left side: 0

Total charge on right side: 16(+1) = +16

Add 16 electrons to right side:

$$8\ H_2S \rightarrow S_8 + 16\ H^+ + 16\ e^-$$

4. B is correct.

An electrolytic cell is a nonspontaneous electrochemical cell that requires the supply of electrical energy (e.g., battery) to initiate the reaction.

The anode is positive, and the cathode is the negative electrode.

For electrolytic and galvanic cells, oxidation occurs at the anode, while reduction occurs at the cathode.

Therefore, Co metal is produced at the cathode because it reduces the product (from an oxidation number of +3 on the left to 0 on the right). Co will not be produced at the anode.

5. C is correct.

The anode in galvanic cells attracts anions.

Anions in solution flow toward the anode, while cations flow toward the cathode.

Oxidation (i.e., loss of electrons) occurs at the anode.

Positive ions are formed while negative ions are consumed at the anode.

Therefore, negative ions flow toward the anode to equalize the charge.

Copyright © Sterling Education.

6. E is correct.

By convention, the reference standard for potential is hydrogen reduction.

7. C is correct.

Calculate the oxidation numbers of species involved and look for the oxidized species (increase in oxidation number).

Cd's oxidation number increases from 0 on the left side of the reaction to +2 on the right.

8. D is correct.

The oxidation number of Fe increases from 0 on the left to +3 on the right.

9. D is correct.

Salt bridge contains cations (positive ions) and anions (negative ions).

Anions flow towards the oxidation half-cell because the oxidation product is positively charged, and the anions are required to balance the charges within the cell.

The opposite is true for cations; they flow towards the reduction half-cell.

10. E is correct.

A redox reaction, or oxidation-reduction reaction, involves the transfer of electrons between two reacting substances.

An oxidation reaction refers explicitly to a substance that is losing electrons, and a reduction reaction refers explicitly to a substance that is gaining reactions.

The oxidation and reduction reactions alone are half-reactions because they occur together to form a whole reaction.

The key to this question is about the *transfer* of electrons between *two* species, referring to the whole redox reaction in its entirety, not just one half-reaction.

An electrochemical reaction occurs during the passage of electric current and involves redox reactions. However, it is not the correct answer to the question posed.

Any duplication (copies, uploads, PDFs) is illegal.

11. A is correct.

The cell reaction is:

$Co\ (s) + Cu^{2+}\ (aq) \rightarrow Co^{2+}\ (aq) + Cu\ (s)$

Separate it into half reactions:

$Cu^{2+}\ (aq) \rightarrow Cu\ (s)$

$Co\ (s) \rightarrow Co^{2+}\ (aq)$

The potentials provided are written in this format:

$Cu^{2+}\ (aq)\ |\ Cu\ (s)$

+0.34 V

It means that for the reduction reaction:

$Cu^{2+}\ (aq) \rightarrow Cu\ (s)$, the potential is +0.34 V

The reverse reaction or oxidation reaction is:

$Cu\ (s) \rightarrow Cu^{2+}\ (aq)$ has opposing potential value: –0.34 V

Obtain the potential values for both half-reactions.

Reverse sign for potential values of oxidation reactions:

$Cu^{2+}\ (aq) \rightarrow Cu\ (s) = +0.34$ V (reduction)

$Co\ (s) \rightarrow Co^{2+}\ (aq) = +0.28$ V (oxidation)

Determine the standard cell potential:

Standard cell potential = sum of half-reaction potential

Standard cell potential = 0.34 V + 0.28 V

Standard cell potential = 0.62 V

Copyright © Sterling Education.

12. E is correct.

The other methods listed involve two or more energy conversions between light and electricity.

It is an electrical device that converts light energy directly into electricity by the photovoltaic effect (i.e., chemical and physical processes).

The operation of a photovoltaic (PV) cell has the following requirements:

1) Light is absorbed, which excites electrons.

2) Separation of charge carries opposite types.

3) The separated charges are transferred to an external circuit.

In contrast, solar panels supply heat by absorbing sunlight.

A photoelectrolytic / photoelectrochemical cell refers either to a type of photovoltaic cell or to a device that splits water into hydrogen and oxygen using only solar illumination.

13. A is correct.

Each half-cell contains an electrode; two half-cells are required to complete a reaction.

14. B is correct.

A: electrolysis can be performed on any metal, not only iron.

C: electrolysis will not boil water nor raise the ship.

D: it is unlikely that the gases would stay in the compartments.

E: electrolysis will not reduce the weight enough to float the ship.

15. C is correct.

Balanced reaction:

$$Zn\ (s) + CuSO_4\ (aq) \rightarrow Cu\ (s) + ZnSO_4\ (aq)$$

Zn is a stronger reducing agent (more likely to be oxidized) than Cu. This can be determined by each element's standard electrode potential (E°).

16. E is correct.

The positive cell potential indicates that the reaction is spontaneous, which favors the formation of products.

Cells that generate electricity spontaneously are considered galvanic cells.

Any duplication (copies, uploads, PDFs) is illegal.

17. B is correct.

A joule is a unit of energy.

18. C is correct.

Electrolysis of aqueous sodium chloride yields hydrogen and chlorine, with aqueous sodium hydroxide remaining in the solution.

Sodium hydroxide is a strong base, which means the solution will be basic.

19. C is correct.

Balanced equation:

$Ag^+ + e^- \rightarrow Ag\ (s)$

Formula to calculate deposit mass:

mass of deposit = (atomic mass × current × time) / 96,500 C

mass of deposit = [107.86 g × 3.50 A × (12 min × 60 s/min)] / 96,500 C

mass of deposit = 2.82 g

20. E is correct.

The anode is the electrode where oxidation occurs.

A salt bridge provides electrical contact between the half-cells.

The cathode is the electrode where reduction occurs.

A spontaneous electrochemical cell is called a galvanic cell.

Copyright © Sterling Education.

Practice Set 2: Questions 21–40

21. A is correct.

Ionization energy (IE) is required to release one electron from an element.

Higher IE means the element is more stable.

Elements with high IE usually need one or two more electrons to achieve stable configuration (e.g., complete valence shell – 8 electrons, such as the noble gases, or 2 electrons as in H).

These elements are more likely to gain an electron and reach stability than lose an electron. When an atom gains electrons, its oxidation number decreases; it is reduced.

Therefore, elements with high IE are easily reduced.

In oxidation-reduction reactions, species that undergo reduction are oxidizing agents because their presence allows the other reactant to be oxidized. Because elements with high IE are easily reduced, they are strong oxidizing agents.

Reducing agents are species that undergo oxidation (i.e., lose electrons).

Elements with high IE do not undergo oxidation readily and therefore are weak reducing agents.

22. D is correct.

Balancing a half-reaction in basic conditions:

The first few steps are identical to balancing reactions in acidic conditions.

Step 1: Balance all atoms except for H and O

$C_8H_{10} \rightarrow C_8H_4O_4^{2-}$ C is already balanced

Step 2: To balance oxygen, add H_2O to the side with fewer oxygen atoms

$$C_8H_{10} + 4\ H_2O \rightarrow C_8H_4O_4^{2-}$$

Step 3: To balance hydrogen, add H^+ to the opposing side of H_2O added in the previous step

$$C_8H_{10} + 4\ H_2O \rightarrow C_8H_4O_4^{2-} + 14\ H^+$$

This next step is the unique additional step for basic conditions.

Step 4: Add equal amounts of OH^- on both sides. The number of OH^- should match the number of H^+ ions. Combine H^+ and OH^- on the same side to form H_2O. If H_2O molecules are on both sides, subtract accordingly to end up with H_2O on one side only.

continued...

Any duplication (copies, uploads, PDFs) is illegal.

There are 14 H^+ ions on the right, so add 14 OH^- ions on both sides:

$$C_8H_{10} + 4\ H_2O + 14\ OH^- \rightarrow C_8H_4O_4^{2-} + 14\ H^+ + 14\ OH^-$$

Combine H^+ and OH^- ions to form H_2O:

$$C_8H_{10} + 4\ H_2O + 14\ OH^- \rightarrow C_8H_4O_4^{2-} + 14\ H_2O$$

H_2O molecules are on both sides, which cancel, and some H_2O remain on one side:

$$C_8H_{10} + 14\ OH^- \rightarrow C_8H_4O_4^{2-} + 10\ H_2O$$

Step 5: Balance charges by adding electrons to the side with a higher/more positive total charge

Total charge on the left side: $14(-1) = -14$

Total charge on the right side: -2

Add 12 electrons to the right side:

$$C_8H_{10} + 14\ OH^- \rightarrow C_8H_4O_4^{2-} + 10\ H_2O + 12\ e^-$$

23. A is correct.

In electrochemical (i.e., galvanic) cells, oxidation occurs at the anode, and reduction occurs at the cathode.

Therefore, CO_2 is produced at the anode because it is an oxidation product (carbon's oxidation number increases from 0 on the left to +2 on the right).

24. A is correct.

Cu^{2+} is being reduced (i.e., gains electrons), while Sn^{2+} is being oxidized (i.e., loses electrons).

Salt bridge contains cations (positive ions) and anions (negative ions).

Copyright © Sterling Education.

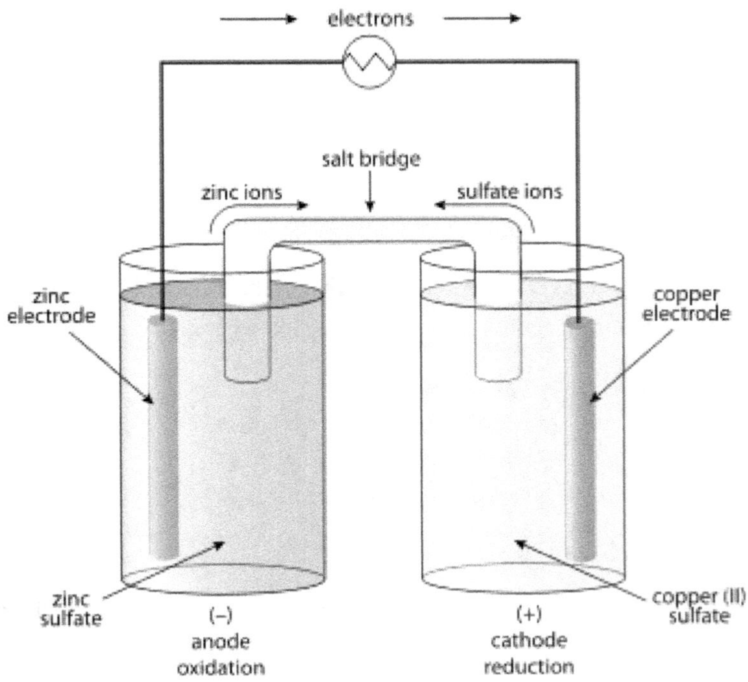

A schematic example of a Zn–Cu galvanic cell

Anions flow towards the oxidation half-cell because the oxidation product is positively charged, and the anions are required to balance the charges within the cell.

The opposite is true for cations; they flow towards the reduction half-cell.

25. B is correct.

In cells, reduction occurs at the cathode, while oxidation occurs at the anode.

26. C is correct.

ZnO is being reduced; the oxidation number of Zn decreases from +2 on the left to 0 on the right.

27. E is correct.

Consider the reactions:

$$Ni\ (s) + Ag^{+}\ (aq) \rightarrow Ag\ (s) + Ni^{2+}\ (aq)$$

Because it is spontaneous, Ni is more likely to be oxidized than Ag.

Ni is oxidized, while Ag is reduced in this reaction.

continued…

Any duplication (copies, uploads, PDFs) is illegal.

Arrange the reactions, so the metal that was oxidized in a reaction is reduced in the following reaction:

$Ni\ (s) + Ag^+\ (aq) \rightarrow Ag\ (s) + Ni^{2+}\ (aq)$

$Cd\ (s) + Ni^{2+}\ (aq) \rightarrow Ni\ (s) + Cd^{2+}\ (aq)$

$Al\ (s) + Cd^{2+}\ (aq) \rightarrow Cd\ (s) + Al^{3+}\ (aq)$

Lastly, $Ag\ (s) + H^+\ (aq) \rightarrow$ no reaction should be first in the order because Ag was not oxidized in this reaction, so it should be before a reaction where Ag is reduced.

$Ag\ (s) + H^+\ (aq) \rightarrow$ no reaction

$Ni\ (s) + Ag^+\ (aq) \rightarrow Ag\ (s) + Ni^{2+}\ (aq)$

$Cd\ (s) + Ni^{2+}\ (aq) \rightarrow Ni\ (s) + Cd^{2+}\ (aq)$

$Al\ (s) + Cd^{2+}\ (aq) \rightarrow Cd\ (s) + Al^{3+}\ (aq)$

The metal being oxidized in the last reaction has the highest tendency to be oxidized.

28. A is correct.

The purpose of the salt bridge is to balance the charges between the two chambers/half-cells.

Oxidation creates cations at the anode, while reduction reduces cations at the cathode.

The ions in the bridge travel to those chambers to balance the charges.

29. B is correct.

Since $G° = –nFE$, when E° is positive, G is negative.

30. C is correct.

The electrons travel through the wires that connect the cells instead of the salt bridge.

The function of the salt bridge is to provide ions to balance charges at the cathode and anode.

31. A is correct.

Electrochemistry is the branch of physical chemistry that studies chemical reactions at the interface of an ionic conductor (i.e., the electrolyte) and an electrode.

Electric charges move between the electrolyte and the electrode through a series of redox reactions, and chemical energy is converted to electrical energy.

Copyright © Sterling Education.

32. B is correct.

Light energy from the sun causes the electron to move towards the silicon wafer. This starts the process of electric generation.

33. B is correct.

Oxidation number (or *oxidation state*) indicates the degree of oxidation (i.e., loss of electrons) in an atom.

If an atom is electron-poor, it has lost electrons and would, therefore, have a positive oxidation number.

If an atom is electron-rich, it has gained electrons and would, therefore, have a negative oxidation number.

34. B is correct.

Reduction potential measures a substance's ability to acquire electrons (i.e., undergo reduction). The more positive the reduction potential, the more likely the substance will be reduced. Reduction potential is generally measured in volts.

35. D is correct.

Batteries run down and need to be recharged, while fuel cells do not run down because they can be refueled.

36. A is correct.

Disproportionation reaction is a reaction where a species undergoes oxidation and reduction in the same reaction.

The first step in balancing this reaction is to write the substance undergoing disproportionation twice on the reactant side.

37. A is correct.

The terms spontaneous electrochemical, galvanic, and voltaic are synonymous.

An example of a voltaic cell is an alkaline battery – it generates electricity spontaneously.

Electrons flow from the anode (oxidation half-cell) to the cathode (reduction half-cell).

Any duplication (copies, uploads, PDFs) is illegal.

38. B is correct.

Reaction at the anode:

$$2\ H_2O \rightarrow O_2 + 4\ H^+ + 4\ e^-$$

Oxygen gas is released, and H^+ ions are added to the solution, which causes the solution to become acidic (i.e., lowers the pH).

39. D is correct.

Nonspontaneous electrochemical cells are electrolysis cells.

In electrolysis, the metal being reduced is produced at the cathode.

40. B is correct.

Electrolysis is nonspontaneous because it needs an electric current from an external source to occur.

Copyright © Sterling Education.

Practice Set 3: Questions 41–60

---

41. C is correct.

Electronegativity indicates the tendency of an atom to attract electrons.

An atom that attracts electrons strongly would be more likely to pull electrons from another atom.

As a result, this atom would be a strong oxidizer because it would cause other atoms to be oxidized when the electrons are pulled towards the atom with high electronegativity.

A strong oxidizing agent is a weak reducing agent – they are opposing attributes.

42. E is correct.

MnO_2 is being reduced into Mn_2O_3.

The oxidation number increases from +4 on the left to +3 on the right.

43. C is correct.

Electrolytic cell: needs electrical energy input

Battery: spontaneously produces electrical energy. A dry cell is a type of battery.

Half-cell: does not generate energy by itself.

44. B is correct.

The anode in galvanic cells attracts anions.

Anions in solution flow toward the anode, while cations flow toward the cathode. Oxidation (i.e., loss of electrons) occurs at the anode.

Positive ions are formed, while negative ions are consumed at the anode.

Therefore, negative ions flow toward the anode to equalize the charge.

45. B is correct.

In electrochemical (i.e., galvanic) cells, oxidation occurs at the anode, and reduction occurs at the cathode.

Therefore, Al metal is produced at the cathode because it is a reduction product (from an oxidation number of +3 on the left to 0 on the right).

46. B is correct.

Anions in solution flow toward the anode, while cations flow toward the cathode.

Oxidation is the loss of electrons. Sodium is a group I element and therefore has a single valence electron. During oxidation, the Na becomes Na^+ with a complete octet.

47. A is correct.

A salt bridge contains cations (positive ions) and anions (negative ions).

Anions flow towards the oxidation half-cell because the oxidation product is positively charged, and the anions are required to balance the charges within the cell.

The opposite is true for cations; they flow towards the reduction half-cell.

Anions in the salt bridge should flow from Cd to Zn half-cell.

48. B is correct.

This can be determined using the electrochemical series:

Equilibrium	**E°**
Li^+ (*aq*) + e^- ↔ Li (*s*)	–3.03 volts
K^+ (*aq*) + e^- ↔ K (*s*)	–2.92
*Ca^{2+} (*aq*) + 2 e^- ↔ Ca (*s*)	–2.87
*Na^+ (*aq*) + e^- ↔ Na (s)	–2.71
Mg^{2+} (*aq*) + 2 e^- ↔ Mg (s)	–2.37
Al^{3+} (*aq*) + 3 e^- ↔ Al (*s*)	–1.66
Zn^{2+} (*aq*) + 2 e^- ↔ Zn (*s*)	–0.76
Fe^{2+} (*aq*) + 2 e^- ↔ Fe (*s*)	–0.44
Pb^{2+} (*aq*) + 2 e^- ↔ Pb (*s*)	–0.13
2 H^+ (*aq*) + 2 e^- ↔ H_2 (*g*)	0.0
Cu^{2+} (*aq*) + 2 e^- ↔ Cu (*s*)	+0.34
Ag^+ (*aq*) + e^- ↔ Ag (*s*)	+0.80
Au^{3+} (*aq*) + 3 e^- ↔ Au (*s*)	+1.50

Copyright © Sterling Education.

For a substance to act as an oxidizing agent for another substance, the oxidizing agent must be located below the substance being oxidized in the series.

Cu is being oxidized, which means only Ag or Au can oxidize Cu.

49. E is correct.

Anions in solution flow toward the anode, while cations flow toward the cathode.

Oxidation (i.e., loss of electrons) occurs at the anode.

Positive ions are formed while negative ions are consumed at the anode.

Therefore, negative ions flow toward the anode to equalize the charge.

50. B is correct.

Electrochemistry is the branch of physical chemistry that studies chemical reactions at the interface of an ionic conductor (i.e., the electrolyte) and an electrode.

Electric charges move between the electrolyte and the electrode through a series of redox reactions, and chemical energy is converted to electrical energy.

51. E is correct.

E° tends to be negative and G positive, because electrolytic cells are nonspontaneous.

Electrons must be forced into the system for the reaction to proceed.

52. C is correct.

To create chlorine gas (Cl_2) from chloride ions (Cl^-), the ion must be oxidized.

In electrolysis, oxidation occurs at the anode, the positive electrode.

The positively-charged electrode would absorb electrons from the ion and transfer them to the cathode for reduction.

53. B is correct.

In electrolytic cells:

The anode is a positively charged electrode where oxidation occurs.

The cathode is a negatively charged electrode where reduction occurs.

54. C is correct.

Nonspontaneous electrochemical cells are electrolysis cells.

In electrolysis, the metal that is being oxidized dissolves at the anode.

Any duplication (copies, uploads, PDFs) is illegal.

55. D is correct.

Dry-cell batteries are the common disposable household batteries. They are based on zinc and manganese dioxide cells.

56. C is correct.

Disproportionation is a redox reaction in which a species is simultaneously reduced and oxidized to form two different products.

Unbalanced reaction: $HNO_2 \rightarrow NO + HNO_3$

Balanced reaction: $3\ HNO_2 \rightarrow 2\ NO + HNO_3 + H_2O$

$2\ H_2O \rightarrow 2\ H_2 + O_2$: decomposition

$H_2SO_3 \rightarrow H_2O + SO_2$: decomposition

$Mg + H_2SO_4 \rightarrow MgSO_4 + H_2$: single replacement

57. D is correct.

Electrolysis is the same process in reverse for the chemical process inside a battery.

Electrolysis is often used to separate elements.

58. A is correct.

Fuel cell automobiles are fueled by hydrogen, and the only emission is water (the statement above is reverse of the true statement).

59. A is correct.

Impure copper is oxidized so that it would happen at the anode. Then, pure copper would plate out (i.e., be reduced) on the cathode.

60. B is correct.

Electrolysis is a chemical reaction when electrical energy is passed through a liquid electrolyte.

Electrolysis utilizes direct electric current (DC) to drive an otherwise non-spontaneous chemical reaction. The voltage needed for electrolysis is called the decomposition potential.

Copyright © Sterling Education.

Practice Set 4: Questions 61–80

61. D is correct.

Calculate the oxidation numbers of all species involved and identify the oxidized species (increase in oxidation number).

Zn's oxidation number increases from 0 on the left side of reaction I to +2 on the right.

62. C is correct.

The lack of an aqueous solution makes it a dry cell. An example of a dry cell is an alkaline battery.

63. B is correct.

Half-reaction: $C_2H_6O \rightarrow HC_2H_3O_2$

Balancing half-reaction in acidic conditions:

Step 1: Balance all atoms except for H and O

$C_2H_6O \rightarrow HC_2H_3O_2$ (C is already balanced)

Step 2: To balance oxygen, add H_2O to the side with fewer oxygen atoms

$C_2H_6O + H_2O \rightarrow HC_2H_3O_2$

Step 3: To balance hydrogen, add H^+ to the opposing side of H_2O added in the previous step

$C_2H_6O + H_2O \rightarrow HC_2H_3O_2 + 4\ H^+$

Step 4: Balance charges by adding electrons to the side with a higher/more positive total charge

Total charge on the left side: 0

Total charge on the right side: 4(+1) = +4

Add 4 electrons to the right side: $C_2H_6O + H_2O \rightarrow HC_2H_3O_2 + 4\ H^+ + 4\ e^-$

64. B is correct.

In electrochemical (i.e., galvanic) cells, oxidation occurs at the anode, and reduction occurs at the cathode.

Therefore, CCl_4 will be produced at the anode because it is an oxidation product (carbon's oxidation number increases from 0 on the left to +4 on the right).

Any duplication (copies, uploads, PDFs) is illegal.

65. C is correct.

Battery: spontaneously produces electrical energy. A dry cell is a type of battery.

Half-cell: does not generate energy by itself

Electrolytic cell: needs electrical energy input

66. D is correct.

Anions in solution flow toward the anode, while cations flow toward the cathode.

Oxidation (i.e., loss of electrons) occurs at the anode.

Positive ions are formed, while negative ions are consumed at the anode.

Therefore, negative ions flow toward the anode to equalize the charge.

67. A is correct.

A reducing agent is a reactant that is oxidized (i.e., loses electrons).

Therefore, the best reducing agent is most easily oxidized.

Reversing each of the half-reactions shows that the oxidation of Cr (*s*) has a potential of +0.75 V, which is greater than the potential (+0.13 V) for the oxidation of Sn^{2+} (*aq*).

Thus, Cr (*s*) is a stronger reducing agent.

68. B is correct.

A salt bridge contains cations (positive ions) and anions (negative ions).

Anions flow towards the oxidation half-cell because the oxidation product is positively charged, and the anions are required to balance the charges within the cell.

The opposite is true for cations; they flow toward the reduction half-cell.

Anions in the salt bridge should flow from Cd to Zn half-cell.

69. B is correct.

When zinc is added to HCl, the reaction is: Zn + HCl → $ZnCl_2$ + H_2O, which means that Zn is oxidized into Zn^{2+}.

The number provided in the problem is reduction potential, so to obtain the oxidation potential, flip the reaction:

$$Zn\ (s) \rightarrow Zn^{2+} + 2\ e^-$$

The $E°$ is inverted and becomes +0.76 V. The reaction occurs because the E(is higher than hydrogen's value (which is set to 0).

Copyright © Sterling Education.

70. E is correct.

Calculate the mass of metal deposited in cathode:

Step 1: Calculate total charge using current and time

Q = current × time

Q = 1 A × (10 minutes × 60 s/minute)

Q = 600 A·s = 600 C

Step 2: Calculate moles of electron that has the same amount of charge

moles e^- = Q / 96,500 C/mol

moles e^- = 600 C / 96,500 C/mol

moles e^- = 6.22×10^{-3} mol

Step 3: Calculate moles of metal deposit

Half-reaction of zinc ion reduction:

Zn^{2+} (*aq*) + 2 $e^- \rightarrow$ Zn (*s*)

moles of Zn = (coefficient Zn / coefficient e^-) × moles e^-

moles of Zn = (½) × 6.22×10^{-3} mol

moles of Zn = 3.11×10^{-3} mol

Step 4: Calculate mass of metal deposit

mass Zn = moles Zn × molecular mass of Zn

mass Zn = 3.11×10^{-3} mol × (65 g/mol)

mass Zn = 0.20 g

71. A is correct.

The oxidation number of Mg increases from 0 to +2, which means that Mg loses electrons (i.e., is oxidized).

Conversely, the oxidation number of Cu decreases, which means that it gains electrons (i.e., it is reduced).

Any duplication (copies, uploads, PDFs) is illegal.

72. D is correct.

Electrolysis uses a direct electric current to provide electricity to a nonspontaneous redox process to drive the reaction. The direct electric current must be passed through an ionic substance or solution that contains electrolytes. Electrolysis is often used to separate elements.

73. D is correct.

Oxidation is the loss of electrons, while reduction is the gain of electrons.

74. D is correct.

E° for the cell is always positive.

75. B is correct.

Galvanic cells are spontaneous and generate electrical energy.

76. D is correct.

In electrolytic cells:

The anode is a positively charged electrode where oxidation occurs.

The cathode is a negatively charged electrode where reduction occurs.

77. B is correct.

Balanced equation:

$$Cu^{2+} + 2\ e^{-} \rightarrow Cu\ (s)$$

Formula to calculate deposit mass:

mass of deposit = atomic mass × moles of electron

Since Cu has 2 electrons per atom, multiply the moles by 2:

mass of deposit = atomic mass × (2 × moles of electron)

mass of deposit = atomic mass × (2 × current × time) / 96,500 C

4.00 g = 63.55 g × (2 × 2.50 A × time) / 96,500 C

time = 4,880 s

time = (4,880 s × 1 min/60s × 1 hr/60 min) = 1.36 hr

Copyright © Sterling Education.

78. C is correct.

Alkaline and dry-cell are common disposable batteries and are non-rechargeable.

Fuel cells require fuel that is going to be consumed and are non-rechargeable.

79. E is correct.

Electrolytic cells do not occur spontaneously; the reaction only occurs with external electrical energy.

80. D is correct.

A redox reaction, or oxidation-reduction reaction, involves the transfer of electrcns between two reacting substances.

An oxidation reaction refers to the substance that is losing electrons, and a reduction reaction refers to the substance gaining electrons.

The oxidation and reduction reactions alone are half-reactions because they occur together to form a whole reaction.

Therefore, half-reaction can represent either a separate oxidation process or a reduction process.

Notes for active learning

Copyright © Sterling Education.

Notes for active learning

Any duplication (copies, uploads, PDFs) is illegal.

Notes for active learning

Copyright © Sterling Education.

APPENDIX

Copyright © Sterling Education.

Periodic Table of the Elements

Key:

atomic number **Symbol** name abridged standard atomic weight

1	2	3	4	5	6	7	8	9	10	11	12	13	14	15	16	17	18
1 **H** hydrogen 1.0080 ± 0.0002																	2 **He** helium 4.0026 ± 0.0001
3 **Li** lithium 6.94 ± 0.06	4 **Be** beryllium 9.0122 ± 0.0001											5 **B** boron 10.81 ± 0.02	6 **C** carbon 12.011 ± 0.002	7 **N** nitrogen 14.007 ± 0.001	8 **O** oxygen 15.999 ± 0.001	9 **F** fluorine 18.998 ± 0.001	10 **Ne** neon 20.180 ± 0.001
11 **Na** sodium 22.990 ± 0.001	12 **Mg** magnesium 24.305 ± 0.002											13 **Al** aluminium 26.982 ± 0.001	14 **Si** silicon 28.085 ± 0.001	15 **P** phosphorus 30.974 ± 0.001	16 **S** sulfur 32.06 ± 0.02	17 **Cl** chlorine 35.45 ± 0.01	18 **Ar** argon 39.95 ± 0.16
19 **K** potassium 39.098 ± 0.001	20 **Ca** calcium 40.078 ± 0.004	21 **Sc** scandium 44.956 ± 0.001	22 **Ti** titanium 47.867 ± 0.001	23 **V** vanadium 50.942 ± 0.001	24 **Cr** chromium 51.996 ± 0.001	25 **Mn** manganese 54.938 ± 0.001	26 **Fe** iron 55.845 ± 0.002	27 **Co** cobalt 58.933 ± 0.001	28 **Ni** nickel 58.693 ± 0.001	29 **Cu** copper 63.546 ± 0.003	30 **Zn** zinc 65.38 ± 0.02	31 **Ga** gallium 69.723 ± 0.001	32 **Ge** germanium 72.630 ± 0.008	33 **As** arsenic 74.922 ± 0.001	34 **Se** selenium 78.971 ± 0.008	35 **Br** bromine 79.904 ± 0.003	36 **Kr** krypton 83.798 ± 0.002
37 **Rb** rubidium 85.468 ± 0.001	38 **Sr** strontium 87.62 ± 0.01	39 **Y** yttrium 88.906 ± 0.001	40 **Zr** zirconium 91.224 ± 0.002	41 **Nb** niobium 92.906 ± 0.001	42 **Mo** molybdenum 95.95 ± 0.01	43 **Tc** technetium [97]	44 **Ru** ruthenium 101.07 ± 0.02	45 **Rh** rhodium 102.91 ± 0.01	46 **Pd** palladium 106.42 ± 0.01	47 **Ag** silver 107.87 ± 0.01	48 **Cd** cadmium 112.41 ± 0.01	49 **In** indium 114.82 ± 0.01	50 **Sn** tin 118.71 ± 0.01	51 **Sb** antimony 121.76 ± 0.01	52 **Te** tellurium 127.60 ± 0.03	53 **I** iodine 126.90 ± 0.01	54 **Xe** xenon 131.29 ± 0.01
55 **Cs** caesium 132.91 ± 0.01	56 **Ba** barium 137.33 ± 0.01	57-71 lanthanoids	72 **Hf** hafnium 178.49 ± 0.01	73 **Ta** tantalum 180.95 ± 0.01	74 **W** tungsten 183.84 ± 0.01	75 **Re** rhenium 186.21 ± 0.01	76 **Os** osmium 190.23 ± 0.03	77 **Ir** iridium 192.22 ± 0.01	78 **Pt** platinum 195.08 ± 0.02	79 **Au** gold 196.97 ± 0.01	80 **Hg** mercury 200.59 ± 0.01	81 **Tl** thallium 204.38 ± 0.01	82 **Pb** lead 207.2 ± 1.1	83 **Bi** bismuth 208.98 ± 0.01	84 **Po** polonium [209]	85 **At** astatine [210]	86 **Rn** radon [222]
87 **Fr** francium [223]	88 **Ra** radium [226]	89-103 actinoids	104 **Rf** rutherfordium [267]	105 **Db** dubnium [268]	106 **Sg** seaborgium [269]	107 **Bh** bohrium [270]	108 **Hs** hassium [269]	109 **Mt** meitnerium [277]	110 **Ds** darmstadtium [281]	111 **Rg** roentgenium [282]	112 **Cn** copernicium [285]	113 **Nh** nihonium [286]	114 **Fl** flerovium [290]	115 **Mc** moscovium [290]	116 **Lv** livermorium [293]	117 **Ts** tennessine [294]	118 **Og** oganesson [294]

57 **La** lanthanum 138.91 ± 0.01	58 **Ce** cerium 140.12 ± 0.01	59 **Pr** praseodymium 140.91 ± 0.01	60 **Nd** neodymium 144.24 ± 0.01	61 **Pm** promethium [145]	62 **Sm** samarium 150.36 ± 0.02	63 **Eu** europium 151.96 ± 0.01	64 **Gd** gadolinium 157.25 ± 0.03	65 **Tb** terbium 158.93 ± 0.01	66 **Dy** dysprosium 162.50 ± 0.01	67 **Ho** holmium 164.93 ± 0.01	68 **Er** erbium 167.26 ± 0.01	69 **Tm** thulium 168.93 ± 0.01	70 **Yb** ytterbium 173.05 ± 0.02	71 **Lu** lutetium 174.97 ± 0.01
89 **Ac** actinium [227]	90 **Th** thorium 232.04 ± 0.01	91 **Pa** protactinium 231.04 ± 0.01	92 **U** uranium 238.03 ± 0.01	93 **Np** neptunium [237]	94 **Pu** plutonium [244]	95 **Am** americium [243]	96 **Cm** curium [247]	97 **Bk** berkelium [247]	98 **Cf** californium [251]	99 **Es** einsteinium [252]	100 **Fm** fermium [257]	101 **Md** mendelevium [258]	102 **No** nobelium [259]	103 **Lr** lawrencium [262]

International Union of Pure and Applied Chemistry (IUPAC), 4 May 2022

Any duplication (copies, uploads, PDFs) is illegal.

Notes for active learning

Copyright © Sterling Education.

Common Chemistry Equations

Throughout the test the following symbols have the definitions specified unless otherwise noted.

L, mL	= liter(s), milliliter(s)	mm Hg	= millimeters of mercury
g	= gram(s)	J, kJ	= joule(s), kilojoule(s)
nm	= nanometer(s)	V	= volt(s)
atm	= atmosphere(s)	mol	= mole(s)

ATOMIC STRUCTURE

$E = h\nu$

$c = \lambda\nu$

E = energy

ν = frequency

λ = wavelength

Planck's constant, $h = 6.626 \times 10^{-34}$ J s

Speed of light, $c = 2.998 \times 10^{8}$ m s^{-1}

Avogadro's number = 6.022×10^{23} mol^{-1}

Electron charge, $e = -1.602 \times 10^{-19}$ coulomb

EQUILIBRIUM

$K_c = \dfrac{[C]^c[D]^d}{[A]^a[B]^b}$, where $a\,A + b\,B \rightleftarrows c\,C + d\,D$

$K_p = \dfrac{(P_C)^c(P_D)^d}{(P_A)^a(P_B)^b}$

$K_a = \dfrac{[H^+][A^-]}{[HA]}$

$K_b = \dfrac{[OH^-][HB^+]}{[B]}$

$K_w = [H^+][OH^-] = 1.0 \times 10^{-14}$ at 25°C

$= K_a \times K_b$

$pH = -\log[H^+]$, $pOH = -\log[OH^-]$

$14 = pH + pOH$

$pH = pK_a + \log\dfrac{[A^-]}{[HA]}$

$pK_a = -\log K_a$, $pK_b = -\log K_b$

Equilibrium Constants

K_c (molar concentrations)

K_p (gas pressures)

K_a (weak acid)

K_b (weak base)

K_w (water)

KINETICS

$\ln[A]_t - \ln[A]_0 = -kt$

$\dfrac{1}{[A]_t} - \dfrac{1}{[A]_0} = kt$

$t_{½} = \dfrac{0.693}{k}$

k = rate constant

t = time

$t_{½}$ = half-life

GASES, LIQUIDS, AND SOLUTIONS

$$PV = nRT$$

$$P_A = P_{total} \times X_A, \text{ where } X_A = \frac{\text{moles A}}{\text{total moles}}$$

$$P_{total} = P_A + P_B + P_C + \ldots$$

$$n = \frac{m}{M}$$

$$K = {}^\circ C + 273$$

$$D = \frac{m}{V}$$

$$KE \text{ per molecule} = \frac{1}{2}mv^2$$

$$\text{Molarity, } M = \text{moles of solute per liter of solution}$$

$$A = abc$$

P = pressure
V = volume
T = temperature
n = number of moles
m = mass
M = molar mass
D = density
KE = kinetic energy
v = velocity
A = absorbance
a = molar absorptivity
b = path length
c = concentration

Gas constant, $R = 8.314 \text{ J mol}^{-1} \text{ K}^{-1}$
$= 0.08206 \text{ L atm mol}^{-1} \text{ K}^{-1}$
$= 62.36 \text{ L torr mol}^{-1} \text{ K}^{-1}$
1 atm = 760 mm Hg
= 760 torr
STP = 0.00°C and 10^5 Pa

THERMOCHEMISTRY/ ELECTROCHEMISTRY

$$q = mc\Delta T$$

$$\Delta S^\circ = \sum S^\circ \text{ products} - \sum S^\circ \text{ reactants}$$

$$\Delta H^\circ = \sum \Delta H_f^\circ \text{ products} - \sum \Delta H_f^\circ \text{ reactants}$$

$$\Delta G^\circ = \sum \Delta G_f^\circ \text{ products} - \sum \Delta G_f^\circ \text{ reactants}$$

$$\Delta G^\circ = \Delta H^\circ - T\Delta S^\circ$$
$$= -RT \ln K$$
$$= -nFE^\circ$$

$$I = \frac{q}{t}$$

q = heat
m = mass
c = specific heat capacity
T = temperature
S° = standard entropy
H° = standard enthalpy
G° = standard free energy
n = number of moles
E° = standard reduction potential
I = current (amperes)
q = charge (coulombs)
t = time (seconds)

Faraday's constant, F = 96,485 coulombs per mole of electrons

$$1 \text{ volt} = \frac{1 \text{ joule}}{1 \text{ coulomb}}$$

 Copyright © Sterling Education.

Annotated Glossary of Chemistry Terms

A

Absolute entropy (of a substance) – the increase in the entropy of a substance as it goes from a perfectly ordered crystalline form at 0 K (where its entropy is zero) to the temperature in question.

Absolute zero – the zero point on the absolute temperature scale; –273.15 °C or 0 K; theoretically, the temperature at which molecular motion ceases (i.e., the system does not emit or absorb energy, atoms at rest).

Absorption spectrum – spectrum associated with the absorption of electromagnetic radiation by atoms (or other species), resulting from transitions from lower to higher energy states.

Accuracy – the degree to which a value is close to the actual value; see *precision.*

Acid – a substance that produces H^+ (*aq*) ions in an aqueous solution and gives a pH of less than 7.0; strong acids ionize entirely or almost entirely in dilute aqueous solution; weak acids ionize only slightly. It turns litmus red.

Acid dissociation constant – an equilibrium constant for dissociating a weak acid.

Acid rain – rainwater with a pH of less than 5.7; caused by the gases NO_2 (vehicle exhaust fumes) and SO_2 (from burning fossil fuels) dissolving in the rain. It kills fish, wildlife, and trees and destroys buildings and lakes.

Acidic salt – contains an ionizable hydrogen atom; does not necessarily produce acidic solutions.

Actinides – the fifteen chemical elements that are between actinium (89) and lawrencium (103).

Activated complex – a structure formed by a collision between molecules in which new bonds form.

Activation energy – the amount of energy that reactants must absorb in their ground states to reach the transition state needed for a reaction to occur.

Active metal – a metal with low ionization energy that loses electrons readily to form cations.

Activity (of a component of ideal mixture) – a dimensionless quantity whose magnitude is equal to the molar concentration in an ideal solution; equal to partial pressure in an ideal gas mixture; 1 for pure solids or liquids.

Activity series – a listing of metals (and hydrogen) in order of decreasing activity.

Actual yield – the amount of a specified pure product obtained from a given reaction; see *theoretical yield.*

Addition reaction – a reaction in which two atoms or groups of atoms are added to a molecule, one on each side of a double or triple bond.

Adhesive forces – forces of attraction between a liquid and another surface.

Adsorption – the adhesion of a species onto the surfaces of particles.

Aeration – the mixing of air into a liquid or a solid.

Any duplication (copies, uploads, PDFs) is illegal.

Alcohol – a hydrocarbon derivative containing a ~OH group attached to a carbon atom, not in an aromatic ring.

Alkali metals – the elements of Group IA on the periodic table (e.g., Na, K, Rb).

Alkaline battery – a dry cell in which the electrolyte contains KOH.

Alkaline earth metals – group IIA metals on the periodic table; see *earth metals*.

Allomer – a substance that has a different composition from another but the same crystalline structure.

Allotropes – elements with different structures (therefore different forms), such as carbon (e.g., diamond, graphite, and fullerene).

Allotropic modifications (allotropes) – different forms of the same element in the same physical state.

Alloy – a mixture of metals. For example, bronze is an alloy formed from copper and tin.

Alloying – mixing metals with other substances (usually other metals) to modify their properties.

Alpha (α) particle – a helium nucleus; helium ion with 2+ charge; an assembly of two protons and two neutrons.

Amorphous solid – a non-crystalline solid with no well-defined ordered structure.

Ampere – unit of electrical current; one ampere equals one coulomb per second.

Amphiprotism – the ability of a substance to exhibit amphiprotic by accepting donated protons.

Amphoterism – the ability to react with both acids and bases; to act as either an acid or a base.

Amplitude – the maximum distance that medium particles carrying the wave move from their rest position.

Anion – a negative ion; an atom or group of atoms that has gained one or more electrons.

Anode – in a cathode ray tube, the positive electrode (electrode at which oxidation occurs); the positive side of a dry cell battery or a cell.

Antibonding orbital – a molecular orbital higher in energy than any of the atomic orbitals from which it is derived; lends instability to a molecule or ion when populated with electrons; denoted with star (*) superscript.

Artificial transmutation – an artificially induced nuclear reaction caused by the bombardment of a nucleus with subatomic particles or small nuclei.

Associated ions – short-lived species formed by the collision of dissolved ions of opposite charges.

Atmosphere – a unit of pressure; the pressure supports a column of mercury 760 mm high at 0 °C.

Atom – a chemical element in its smallest form; it comprises neutrons and protons within the nucleus and electrons circling the nucleus; it is the smallest part of an element that can exist.

Atomic mass unit (amu) – one-twelfth of the mass of an atom of the carbon-12 isotope; used for stating atomic and formula weights; known as a dalton.

Atomic number – represents an element corresponding with the number of protons within the nucleus; the number of protons in the nucleus of the atom.

 Copyright © Sterling Education.

Atomic orbital (*AO*) – a region or volume in space where the probability of finding electrons is highest.

Atomic radius – radius of an atom.

Atomic weight – weighted average of the masses of the constituent isotopes of an element; the relative masses of atoms of different elements.

Aufbau (or *building up*) principle – describes the order in which electrons fill orbitals in atoms.

Autoionization – an ionization reaction between identical molecules.

Avogadro's Law – equal volumes of gases contain the same number of molecules at the same temperature and pressure.

Avogadro's number (N_A) – the number (6.022×10^{23}) of atoms, molecules, or particles found in precisely 1 mole of a substance.

B

Background radiation – extraneous to an experiment; usually the low-level natural radiation from cosmic rays and trace radioactive substances present in the environment.

Band – a series of very closely spaced nearly continuous molecular orbitals that belong to the crystal as a whole.

Band of stability – band containing nonradioactive nuclides in a plot of neutrons *vs.* their atomic number.

Band theory of metals – the theory that accounts for the bonding and properties of metallic solids.

Barometer – a device used to measure the pressure in the atmosphere.

Base – a substance that produces ^-OH (*aq*) ions in an aqueous solution; accepts a proton and has a high pH; strongly soluble bases are soluble in water and are entirely dissociated; weak bases ionize only slightly; a typical example of a base is sodium hydroxide (NaOH). It turns litmus blue.

Basic anhydride – the oxide of a metal that reacts with water to form a base.

Basic salt – a salt containing an ionizable OH group.

Beta (β) particle – an electron emitted from the nucleus when a neutron decays to a proton and an electron.

Binary acid – a binary compound in which H is bonded to one or more electronegative nonmetals.

Binary compound – consists of two elements; it may be ionic or covalent.

Binding energy (nuclear binding energy) – the energy equivalent ($E = mc^2$) of the mass deficiency of an atom (where E is the energy in joules, m is the mass in kilograms, and c is the speed of light in m/s^2).

Boiling – the phase transition of a liquid vaporizing.

Boiling point – the temperature at which the vapor pressure of a liquid is equal to the applied pressure; the *condensation point*.

Boiling point elevation – the increase in the boiling point of a solvent caused by the dissolution of a nonvolatile solute.

Any duplication (copies, uploads, PDFs) is illegal.

Bomb calorimeter – a device used to measure the heat transfer between a system and its surroundings at constant volume.

Bond – the attraction and repulsion between atoms and molecules is a cornerstone of chemistry.

Bond energy – the amount of energy necessary to break one mole of bonds in a substance, dissociating the substance in its gaseous state into atoms of its elements in the gaseous state.

Bond order – half the number of electrons in bonding orbitals minus half the electrons in antibonding orbitals.

Bonding orbital – a molecular orbit lower in energy than any of the atomic orbitals from which it is derived; lends stability to a molecule or ion when populated with electrons.

Bonding pair – pair of electrons involved in a covalent bond.

Boron hydrides – binary compounds of boron and hydrogen.

Born-Haber cycle – a series of reactions (and the accompanying enthalpy changes) which, when summed, represent the hypothetical one-step reaction by which elements in their standard states are converted into crystals of ionic compounds (and the accompanying enthalpy changes).

Boyle's Law – at a constant temperature, the volume occupied by a definite mass of a gas is inversely proportional to the applied pressure.

Breeder reactor – a nuclear reactor that produces more fissionable nuclear fuel than it consumes.

Brønsted-Lowrey acid – a chemical species that donates a proton.

Brønsted-Lowrey base – a chemical species that accepts a proton.

Buffer solution – resists change in pH; contains either a weak acid and a soluble ionic salt of the acid or a weak base and a soluble ionic salt of the base.

Buret – a piece of volumetric glassware, usually graduated in 0.1 mL intervals, used to deliver solutions for titrations in a quantitative (drop-like) manner; also spelled *burette*.

C

Calorie – the amount of heat required to raise the temperature of one gram of water from 14.5 °C to 15.5 °C; 1 calorie = 4.184 joules.

Calorimeter – a device used to measure the heat transfer between a system and its surroundings.

Canal ray – a stream of positively charged particles (cations) that moves toward the negative electrode in cathode ray tubes; observed to pass through canals in the negative electrode.

Capillary – a tube having a very small inside diameter.

Capillary action – the drawing of a liquid up the inside of a small-bore tube when adhesive forces exceed cohesive forces; the depression of the surface of the liquid when cohesive forces exceed the adhesive forces.

Catalyst – a chemical compound used to change the rate (to speed or slow it) of a regenerated reaction (i.e., not consumed) at the end of the reaction.

Copyright © Sterling Education.

Catenation – the bonding of atoms of the same element into chains or rings (i.e., the ability of an element to bond with itself).

Cathode – the electrode at which reduction occurs; in a cathode ray tube, the negative electrode.

Cathodic protection – protection of a metal (making a cathode) against corrosion by attaching it to a sacrificial anode of a more easily oxidized metal.

Cathode ray tube – a closed glass tube containing gas under low pressure, with electrodes near the ends and a luminescent screen near the positive electrode; produces cathode rays when a high voltage is applied.

Cation – a positive ion; an atom or group of atoms that has lost one or more electrons.

Cell potential – the potential difference, E_{cell}, between oxidation and reduction half-cells under nonstandard conditions; the force in a galvanic cell pulls electrons through a reducing agent to an oxidizing agent.

Central atom – an atom in a molecule or polyatomic ion bonded to more than one other atom.

Chain reaction – a reaction that, once initiated, sustains itself and expands; a reaction in which reactive species, such as radicals, are produced in more than one step, allowing these reactive species to propagate the chain reaction.

Charles' Law – at constant pressure, the volume occupied by a definite mass of gas is directly proportional to its absolute temperature.

Chemical bonds – the attractive forces holding atoms together in elements or compounds.

Chemical change – when one or more new substances are formed.

Chemical equation – description of a chemical reaction by placing the formulas of the reactants on the left of an arrow and the formulas of the products on the right.

Chemical equilibrium – a state of dynamic balance in which the rates of forward and reverse reactions are equal; there is no net change in concentrations of reactants or products while a system is at equilibrium.

Chemical kinetics – studies rates and mechanisms of chemical reactions and factors they depend on.

Chemical periodicity – the variations in properties of elements with their position in the periodic table.

Chemical reaction – the change of one or more substances into another or multiple substances.

Cloud chamber – a device for observing the paths of speeding particles as vapor molecules condense on them to form fog-like tracks.

Cobalt chloride paper – water test; water changes the color from blue to pink.

Coefficient of expansion – the ratio of the change in the length or the volume of a body to the original length or volume for a unit change in temperature.

Cohesive forces – the forces of attraction among particles of a liquid.

Colligative properties – physical properties of solutions that depend upon the number but not the kind of solute particles present.

Any duplication (copies, uploads, PDFs) is illegal.

Collision theory – the theory of reaction rates that states that effective collisions between reactant molecules must occur for the reaction to occur.

Colloid – a heterogeneous mixture in which solute-like particles do not settle (e.g., many kinds of milk).

Combination reaction – two substances (elements or compounds) combine to form one compound.

Combustible – classification of liquid substances that burn based on flashpoints; any liquid having a flashpoint at or above 37.8 °C (100 °F) but below 93.3 °C (200 °F), except any mixture having components with flashpoints of 93.3 °C (200 °F) or higher, the total of which makes up 99% or more of the volume of the mixture.

Combustion (or *burning*) – an exothermic reaction between an oxidant and fuel with heat and often light.

Common ion effect – suppression of ionization of a weak electrolyte by the presence in the same solution of a strong electrolyte containing one of the same ions as the weak electrolyte.

Complex ions – ions resulting from coordinating covalent bonds between simple ions and other ions or molecules.

Composition stoichiometry – describes the quantitative (mass) relationships among elements in compounds.

Compound – a substance of two or more chemically bonded elements in fixed proportions; can be decomposed into constituent elements.

Compressed gas – a single or mixture of gases having (in a container) an absolute pressure exceeding 40 psi at 21.1 °C (70 °F).

Compression – an area in a longitudinal wave where the particles are closer and pushed in.

Concentration – the amount of solute per unit volume, the mass of solvent or solution.

Condensation – the phase change from gas to liquid.

Condensed phases – the liquid and solid phases; phases in which particles interact strongly.

Condensed states – the solid and liquid states.

Conduction – heat transfer between substances in direct contact with each other (i.e., must be touching); when particles of a hotter substance vibrate, these molecules bump into nearby particles and transfer some energy.

Conduction band – a vacant or partially filled band of energy levels just higher in energy than a filled band; a band within which, or into which, electrons must be promoted to allow electrical conduction to occur in a solid.

Conductor – a material that allows electric flow more freely.

Conjugate acid-base pair – in Brønsted-Lowry terms, reactant and product that differ by a proton, H^+.

Conformations – structures of a compound that differ by the extent of rotation about a single bond.

Continuous spectrum – contains wavelengths in a specified region of the electromagnetic spectrum.

Control rods – rods of materials such as cadmium or boron steel that act as neutron absorbers (not merely moderators), used in nuclear reactors to control neutron fluxes and therefore fission rates.

 Copyright © Sterling Education.

Conjugated double bonds – double bonds separated from each other by one single bond –C=C–C=C–

Contact process – the industrial process for sulfur trioxide (SO_3) and sulfuric acid (H_2SO_4) production from sulfur dioxide (SO_2).

Convection – the physical flow of matter when heat flows by energized molecules from one place to another through the movement of fluids. Transfer of heat through a liquid or a gas occurs when molecules of the liquid or gas move and carry the heat.

Coordinate covalent bond – a covalent bond with shared electrons furnished by the same species; a bond between a Lewis acid and a Lewis base.

Coordination compound or complex – a compound containing coordinate covalent bonds.

Coordination number – the number of donor atoms coordinated to a metal; the number of nearest neighbors of an atom or ion in describing crystals.

Coordination sphere – the metal ion and its coordinating ligands, but no uncoordinated counter-ions.

Corrosion – oxidation of metals (e.g., rusting) in the presence of air and moisture

Coulomb – the SI unit of electrical charge; unit symbol – C.

Covalent bond – a force of attraction (chemical bond) formed by the sharing of electron pairs between two atoms.

Covalent compounds – compounds made of two or more nonmetal atoms bonded by sharing valence electrons.

Critical mass – the minimum mass of a particular fissionable nuclide in a given volume required to sustain a nuclear chain reaction.

Critical point – the combination of critical temperature and critical pressure of a substance.

Critical pressure – the pressure required to liquefy a gas (vapor) at its *Critical temperature*.

Critical temperature – the temperature above which a gas cannot be liquefied; the temperature above which a substance cannot exhibit distinct gas and liquid phases.

Crystal – a solid packed with ions, molecules, or atoms in an orderly fashion.

Crystal field stabilization energy – a measure of the net energy of stabilization gained by a metal ion's nonbonding d electrons due to complex formation.

Crystal field theory – bonding in transition metal complexes in which ligands and metal ions are treated as point charges; a purely ionic model; ligand point charges represent the crystal (electrical) field perturbing the metal's d orbitals containing nonbonding electrons.

Crystal lattice – a pattern of arrangement of particles in a crystal.

Crystal lattice energy – the amount of energy that holds a crystal together; the energy change when a mole of solid forms from its constituent molecules or ions (for ionic compounds) in their gaseous state (consistently negative).

Crystalline solid – a solid characterized by a regular, ordered arrangement of particles.

Any duplication (copies, uploads, PDFs) is illegal.

Curie (Ci) – the basic unit to describe the intensity of radioactivity in a sample of material; one curie equals 37 billion disintegrations per second or approximately the amount of radioactivity given off by 1 gram of radium.

Current – a flow of charged particles, such as electrons or ions, moving through an electrical space or conductor. It is measured as the net rate of flow of electric charge; the unit is an Ampere (A).

Cuvette – glassware used in spectroscopic experiments; usually made of plastic, glass, or quartz, and should be as clean and transparent as possible.

Cyclotron – a device for accelerating charged particles along a spiral path.

D

Dalton's Law (or the *law of partial pressures*) – the pressure exerted by a mixture of gases is the sum of the partial pressures of the individual gases.

Daughter nuclide – nuclide produced in nuclear decay.

Debye – the unit used to express dipole moments.

Degenerate – in orbitals, describes orbitals of the same energy.

Deionization – the removal of ions; in the case of water, mineral ions such as sodium, iron, and calcium.

Deliquescence – substances that absorb water from the atmosphere to form liquid solutions.

Delocalization – in reference to electrons, bonding electrons distributed among more than two atoms bonded; occurs in species that exhibit resonance.

Density – mass per unit volume; $D = M \times V$.

Deposition – settling particles within a solution; the direct solidification of vapor by cooling; see *sublimation.*

Derivative – a compound that can be imagined arising from a parent compound by replacing one atom with another atom or group of atoms; used extensively in organic chemistry to identify compounds.

Detergent – a soap-like emulsifier with sulfate, SO_3, or a phosphate group instead of carboxylate group.

Deuterium – an isotope of hydrogen whose atoms are twice as massive as ordinary hydrogen; deuterium atoms contain a proton and a neutron in the nucleus.

Dextrorotatory – refers to an optically active substance that rotates plane-polarized light clockwise, also known as *dextro.*

Diagonal similarities – chemical similarities in the Periodic Table of Elements of Period 2 to elements of Period 3 one group to the right, especially evident toward the left of the periodic table.

Diamagnetism – weak repulsion by a magnetic field.

Differential Scanning Calorimetry (DSC) – a technique for measuring temperature, direction, and magnitude of thermal transitions in a sample material by heating/cooling and comparing the amount of energy required to maintain its rate of temperature increase or decrease with an inert reference material under similar conditions.

 Copyright © Sterling Education.

Differential Thermal Analysis (DTA) – a technique for observing the temperature, direction, and magnitude of thermally induced transitions in a material by heating/cooling a sample and comparing its temperature with an inert reference material under similar conditions.

Differential thermometer – a thermometer used to measure minimal temperature changes accurately.

Dilution – the process of reducing the concentration of a solute in a solution, usually by mixing with more solvent.

Dimer – a molecule formed by combining two smaller (identical) molecules.

Dipole – electric or magnetic separation of charge; charge separation between two covalently bonded atoms.

Dipole-dipole interactions – attractive electrostatic forces between polar molecules (i.e., between molecules with permanent dipoles).

Dipole moment – the product of the distance separating opposite charges of equal magnitude; a measure of the polarity of a bond or molecule; a measured dipole refers to the dipole moment of an entire molecule.

Dispersing medium – the solvent-like phase in a colloid.

Dispersed phase – the solute-like species in a colloid.

Displacement reactions – reactions in which one element displaces another from a compound.

Disproportionation reactions – redox reactions in which the oxidizing agent and the reducing agent are the same species.

Dissociation – in an aqueous solution, the process by which a solid ionic compound separates into its ions

Dissociation constant – equilibrium constant for dissociating a complex ion into a simple ion and coordinating species (ligands).

Dissolution or solvation – the spread of ions in a monosaccharide.

Distilland – the material in a distillation apparatus that is to be distilled.

Distillate – the material in a distillation apparatus collected in the receiver.

Distillation – separating a liquid mixture into its components based on differences in boiling points; the process in which components of a mixture are separated by boiling away the more volatile liquid; the vaporization of a liquid by heating and then the condensation of the vapor by cooling.

Domain – a cluster of atoms in a ferromagnetic substance, which align in the same direction in the presence of an external magnetic field.

Donor atom – a ligand atom whose electrons are shared with a Lewis acid.

***d*-orbitals** – beginning in the third energy level, a set of five degenerate orbitals per energy level, higher in energy than *s* and *p* orbitals of the same energy level.

Dosimeter – a small, calibrated electroscope worn by laboratory personnel to measure incident ionizing radiation or chemical exposure.

Double bond – covalent bond resulting from the sharing of four electrons (two pairs) between two atoms.

Any duplication (copies, uploads, PDFs) is illegal.

Double salt – a solid consisting of two co-crystallized salts.

Doublet – two peaks or bands of about equal intensity appearing close on a spectrogram.

Downs cell – an electrolytic cell for the commercial electrolysis of molten sodium chloride.

DP number – the degree of polymerization; the average number of monomer units per polymer unit.

Dry cells (voltaic cells) – ordinary batteries for appliances (e.g., flashlights, radios).

Dumas method – a method used to determine the molecular weights of volatile liquids.

Dynamic equilibrium – an equilibrium in which the processes occur continuously with no net change.

E

Earth metal – highly reactive elements in group IIA of the periodic table (includes beryllium, magnesium, calcium, strontium, barium, and radium); see *alkaline earth metal.*

Effective collisions – a collision between molecules resulting in a reaction, one in which the molecules collide with proper relative orientations and sufficient energy to react.

Effective molality – the sum of the molalities of solute particles in a solution.

Effective nuclear charge – the nuclear charge experienced by the outermost electrons of an atom; the actual nuclear charge minus the effects of shielding due to inner-shell electrons (e.g., a set of dx_2-y_2 and dz_2 orbitals); those d orbitals within a set with lobes directed along the *x*, *y,* and *z* axes.

Electrical conductivity – the measure of how easily an electric current can flow through a substance.

Electric charge – a measured property (coulombs) that determines electromagnetic interaction.

Electrochemical cell – using a chemical reaction, an electromotive force is generated.

Electrochemistry – the study of chemical changes produced by electrical current and electricity production by chemical reactions.

Electrodes – surfaces upon which oxidation and reduction half-reactions occur in electrochemical cells; a conductor dips into an electrolyte and allows the electrons to flow to and from the electrolyte.

Electrode potentials – potentials, E, of half-reactions as reductions vs. standard hydrogen electrode.

Electrolysis – 1) occurs in electrolytic cells; chemical decomposition occurs by passing an electric current through a solution containing ions. 2) producing a chemical change using electricity; used to split up water into H and O_2.

Electrolyte – a substance (i.e., anions, cations) which, when dissolved in water, conducts electricity. An ionic solution that conducts a certain amount of current is categorized into two types: weak and strong.

Electrolytic cells – electrochemical cells in which electrical energy causes nonspontaneous redox reactions to occur (i.e., forced to occur by applying an outside source of electrical energy).

Electrolytic conduction – electrical current passes through ions in a solution or pure liquid.

Electromagnetic radiation – energy propagated using electric and magnetic fields that oscillate in directions perpendicular to the direction of travel of the energy; a type of wave that can go through vacuums as well as material; classified as a “self-propagating wave.”

 Copyright © Sterling Education.

Electromagnetism – fields of an electric charge and electric properties that change how particles move and interact.

Electromotive force – a device that gains energy as electric charges pass through it.

Electromotive series – the relative order of tendencies for elements and their simple ions to act as oxidizing or reducing agents; also known as the "activity series."

Electron – a subatomic particle having a mass of 0.00054858 amu and a charge of –1.

Electron affinity – the energy absorbed in the process in which an electron is added to a neutral isolated gaseous atom to form a gaseous ion with a 1– charge; it has a negative value if energy is released.

Electron configuration – the specific distribution of electrons in atomic orbitals of atoms or ions.

Electron-deficient compounds – at least one atom (other than H) shares fewer than eight electrons.

Electron shells – an orbital around an atom's nucleus with a fixed number of electrons (usually 2 or 8).

Electronic transition – the transfer of an electron from one energy level to another.

Electronegativity – a measure of the relative tendency of an atom to attract electrons to itself when chemically combined with another atom.

Electronic geometry – the geometric arrangement of orbitals containing the shared and unshared electron pairs surrounding the central atom or polyatomic ion.

Electrophile – positively charged or electron-deficient.

Electrophoresis – separating ions by migration rate and direction of migration in an electric field.

Electroplating – a metal is covered with another metal layer using electricity; plating a metal onto a (cathodic) surface by electrolysis.

Element – a substance that cannot be decomposed into simpler substances by chemical means; defined by its *atomic number*. A substance that cannot be split into simpler substances by chemical means.

Eluant (or eluent) – the solvent used in the process of elution, as in liquid chromatography.

Eluate – a solvent (or mobile phase) which passes through a chromatographic column and removes the sample components from the stationary phase.

Emission spectrum – the emission of electromagnetic radiation by atoms (or other species) resulting from electronic transitions from higher to lower energy states.

Empirical formula – the simplest whole-number ratio of atoms of each element present in a compound; also known as the *simplest formula*.

Emulsifying agent – a substance that coats the particles of the dispersed phase and prevents coagulation of colloidal particles; an emulsifier.

Emulsion – colloidal suspension of a liquid in a liquid.

Endothermic – describes processes that absorb heat energy (H).

Endothermicity – the absorption of heat by a system as the process occurs.

Endpoint – the point at which an indicator changes color and titration stops.

Any duplication (copies, uploads, PDFs) is illegal.

Energy – a system's ability to do work.

Enthalpy (*H*) – the heat content of a specific amount of substance; *E*= PV.

Entropy (*S*) – a thermodynamic state or property that measures the degree of disorder (i.e., randomness) of a system; the amount of energy not available for work in a closed thermodynamic system (usually denoted by S).

Enzyme – a protein that acts as a catalyst in biological systems.

Equation of state – an expression that describes the behavior of matter in a given state; the van der Waals equation describes the behavior of the gaseous state.

Equilibrium or chemical equilibrium – a state of dynamic balance with the rates of forward and reverse reactions equal; the state of a system where neither forward nor reverse reaction is thermodynamically favored.

Equilibrium constant – a quantity characterizing equilibrium position for a reversible reaction; its magnitude is equal to mass action expression at equilibrium; equilibrium, "K," varies with temperature.

Equivalence point – the point when chemically equivalent amounts of reactants have reacted.

Equivalent weight – an oxidizing or reducing agent whose mass gains (oxidizing agents) or loses (reducing agents) 6.022×10^{23} electrons in a redox reaction.

Evaporation – vaporization of a liquid below its boiling point.

Evaporation rate – the rate at which a particular substance vaporizes (evaporate) compared to the rate of a known substance, such as ethyl ether, especially useful for health and fire-hazard considerations.

Excited state – any state other than the ground state of an atom or molecule; see *ground state*.

Exothermic – describes processes that release heat energy (*H*).

Exothermicity – the release of heat by a system as a process occurs.

Explosive – a chemical or compound that causes a sudden, almost instantaneous release of pressure, gas, heat, and light when subjected to sudden shock, pressure, high temperature, or applied potential.

Explosive limits – the range of concentrations over which a flammable vapor mixed with the proper ratios of air will ignite or explode if a source of ignition is provided.

Extensive property – a property that depends upon the amount of material in a sample.

Extrapolate – to estimate the value of a result outside the range of a series of known values; a technique used in the standard additions calibration procedure.

F

Faraday constant – a unit of electrical charge widely used in electrochemistry and equal to ~ 96,500 coulombs; represents 1 mole of electrons, or the Avogadro number of electrons: 6.022×10^{23} electrons.

Faraday's law of electrolysis – a two-part law that Michael Faraday published about electrolysis. 1. the mass of a substance altered at an electrode during electrolysis is directly proportional to the quantity of electricity transferred at that electrode. 2. the mass of an elemental material altered at an electrode is directly proportional to the element's equivalent weight; one equivalent weight of a substance is produced at each electrode during the passage of 96,487 coulombs of charge through an electrolytic cell.

 Copyright © Sterling Education.

Fast neutron – a neutron ejected at high kinetic energy in a nuclear reaction.

Ferromagnetism – the ability of a substance to become permanently magnetized by exposure to an external magnetic field.

Flashpoint – the temperature at which a liquid yields enough flammable vapor to ignite; various recognized industrial testing methods exist; therefore, the method used must be specified.

Fluorescence – absorption of high-energy radiation and subsequent emission of visible light.

First Law of Thermodynamics – the amount of energy in the universe is constant (i.e., energy is neither created nor destroyed in ordinary chemical reactions and physical changes); known as the Law of Conservation of Energy.

Fluids – substances that flow freely; gases and liquids.

Flux – a substance added to react with the charge or a product of its reduction; in metallurgy, it is usually added to lower the melting point.

Foam – colloidal suspension of a gas in a liquid.

Formal charge – a method of counting electrons in a covalently bonded molecule or ion; it counts bonding electrons as though they were equally shared between the two atoms.

Formula – a combination of symbols that indicates the chemical composition of a substance.

Formula unit – the smallest repeating unit of a substance; the molecule for nonionic substances.

Formula weight – the mass of one formula unit of a substance in atomic mass units.

Fossil fuels – formed from the remains of plants and animals that lived millions of years ago.

Fractional distillation – when a fractioning column is used in a distillation apparatus to separate the components of a liquid mixture with different boiling points.

Fractional precipitation – removal of some ions from a solution by precipitation while leaving other ions with similar properties in the solution.

Free energy change – the indicator of the spontaneity of a process at constant T and P (e.g., if ΔG is negative, the process is spontaneous).

Free radical – a highly reactive chemical species carrying no charge and having a single unpaired electron in an orbital.

Freezing – phase transition from liquid to solid.

Freezing point depression – the decrease in the freezing point of a solvent caused by the presence of a solute.

Frequency – the number of repeating points on a wave that passes a given observation point per unit time; the unit is 1 hertz = 1 cycle per 1 second.

Fuel – any substance that burns in oxygen to produce heat.

Fuel cells – a voltaic cell that converts the chemical energy of a fuel and an oxidizing agent directly into electrical energy continuously.

Any duplication (copies, uploads, PDFs) is illegal.

G

Gamma (γ) ray – a highly penetrating type of nuclear radiation similar to X-ray radiation, except that it comes from within the nucleus of an atom and has higher energy; energy-wise, very similar to cosmic rays, except that cosmic rays originate from outer space.

Galvanic cell – a battery made from an electrochemical cell with two metals connected by a salt bridge.

Galvanizing – placing a thin layer of zinc on a ferrous material to protect the underlying surface from corrosion.

Gangue – sand, rock, and other impurities surrounding the mineral of interest in an ore.

Gas – a state of matter in which the particles have no definite shape or volume, though they fill their container.

Gay-Lussac's Law – the expression used for each of the two relationships named after the French chemist Joseph Louis Gay-Lussac concerning the properties of gases; more usually applied to his law of combining volumes.

Geiger counter – a gas-filled tube that discharges electrically when ionizing radiation passes through it.

Gel – colloidal suspension of a solid dispersed in a liquid; a semi-rigid solid.

Gibbs (free) energy – the thermodynamic state function of a system that indicates the amount of energy available for the system to do useful work at constant T and P; a value that indicates the spontaneity of a reaction (usually denoted by *G*).

Graham's Law – the rates of effusion of gases are inversely proportional to the square roots of their molecular weights or densities.

Ground state – the lowest energy state or most stable state of an atom, molecule, or ion; see *excited state*.

Group – a vertical column in the periodic table; known as a family.

H

Haber process – a process for the catalyzed industrial production of ammonia from N_2 and H_2 at high temperature and pressure.

Half-cell – the compartment in which the oxidation or reduction half-reaction occurs in a voltaic cell.

Half-life – the time required for half of a reactant to be converted into product(s); the time required for half of a given sample to undergo radioactive decay.

Half-reaction – the oxidation or the reduction part of a redox reaction.

Halogens – group VIIA elements: F, Cl, Br, I; halogens are nonmetals.

Hard water – water high in dissolved minerals that makes it difficult to form lather with soap.

Heat – a form of energy that flows between two samples of matter because of temperature differences.

Heat capacity – the amount of heat required to raise the temperature of a mass one degree Celsius.

Heat of condensation – the amount of heat that must be removed from one gram of vapor at its condensation point to condense the vapor with no change in temperature.

 Copyright © Sterling Education.

Heat of crystallization – the amount of heat that must be removed from one gram of a liquid at its freezing point to freeze it with no change in temperature.

Heat of fusion – the amount of heat required to melt one gram of a solid at its melting point with no change in temperature; usually expressed in J/g; the molar heat of fusion is the amount of heat required to melt one mole of a solid at its melting point with no change in temperature and is usually expressed in kJ/mol.

Heat of solution – the amount of heat absorbed in forming a solution that contains one mole of the solute; the value is positive if heat is absorbed (endothermic) and negative if heat is released (exothermic).

Heat of vaporization – the amount of heat required to vaporize one gram of a liquid at its boiling point with no change in temperature; usually expressed in J/g; the molar heat of vaporization is the amount of heat required to vaporize one mole of liquid at its boiling point with no change in temperature and is usually expressed as ion kJ/mol.

Heisenberg uncertainty principle – states that it is impossible to determine the momentum and position of an electron simultaneously with absolute accuracy.

Henry's Law – the gas pressure above a solution is proportional to concentration of the gas in solution.

Hess' Law of heat summation – the enthalpy change for a reaction is the same whether it occurs in one step or a series of steps.

Heterogeneous catalyst – exists in a different phase (solid, liquid, or gas) from the reactants; a contact catalyst.

Heterogeneous equilibria – equilibria involving species in more than one phase.

Heterogeneous mixture – a mixture that does not have uniform composition and properties throughout.

Heteronuclear – consisting of different elements.

High spin complex – crystal field designation for an outer orbital complex; t_{2g} and e_g orbitals are singly occupied before pairing occurs.

Homogeneous catalyst – in the same phase (solid, liquid, or gas) as the reactants.

Homogeneous equilibria – when *reagents* and *products* are in the same phase (gases, liquids, or solids).

Homogeneous mixture – a mixture which has uniform composition and properties throughout.

Homologous series – compounds with each member differing from the next by a specific number and kind of atoms.

Homonuclear – consisting of only one element.

Hund's rule – single electrons must occupy orbitals of a given sublevel before pairing begins; see *Aufbau* (or *building up*) *principle*.

Hybridization – mixing atomic orbitals to form a new set of atomic orbitals with the same electron capacity and properties and energies intermediate between the original unhybridized orbitals.

Hydrate – a solid compound that contains a definite percentage of bound water.

Hydrate isomers – crystalline complexes that differ in whether water exists inside or outside the coordination sphere.

Any duplication (copies, uploads, PDFs) is illegal.

Hydration – the reaction of a substance with water.

Hydration energy – the energy change accompanying the hydration of a mole of gas and ions.

Hydride – a binary compound of hydrogen.

Hydrocarbons – compounds that contain only carbon and hydrogen.

Hydrogen bond – a relatively strong dipole-dipole interaction (but still considerably weaker than the covalent or ionic bonds) between molecules containing hydrogen directly bonded to a small, highly electronegative atom, such as N, O or F.

Hydrogenation – the reaction in which hydrogen adds across a double or triple bond.

Hydrogen-oxygen fuel cell – hydrogen is the fuel (reducing agent), and oxygen is the oxidizing agent.

Hydrolysis – the reaction of a substance with water or its ions.

Hydrolysis constant – an equilibrium constant for a hydrolysis reaction.

Hydrometer – a device used to measure the densities of liquids and solutions.

Hydrophilic colloids – colloidal particles that repel water molecules.

I

Ideal gas – a hypothetical gas that obeys the postulates of the Kinetic Molecular Theory.

Ideal gas law – the product of pressure and the volume of an ideal gas is directly proportional to the number of moles of the gas and the absolute temperature. $PV = nRT$

Ideal solution – obeys Raoult's Law strictly.

Immiscible liquids – do not mix to form a solution (e.g., oil and water).

Indicators – for acid-base titrations, organic compounds that exhibit different colors in solutions of different acidities, determine the point at which the reaction between two solutes is complete.

Inert pair effect – characteristic of the post-transition minerals; the tendency of the electrons in the outermost atomic *s* orbital to remain un-ionized or unshared in compounds of post-transition metals.

Inhibitory catalyst – an inhibitor; a catalyst that decreases the rate of reaction.

Inner orbital complex – valence bond designation for a complex in which the metal ion utilizes d orbitals for one shell inside the outermost occupied shell in its hybridization.

Inorganic chemistry – a part of chemistry concerned with inorganic (non-carbon-based) compounds.

Insulator – a material that resists the flow of electric current or heat transfer; it does not allow heat to flow easily.

Insoluble compound – a substance that will not dissolve in a solvent, even after mixing.

Integrated rate equation – an expression giving the concentration of a reactant remaining after a specified time; it has a different mathematical form for different orders of reactants.

Intermolecular forces – forces between individual particles (atoms, molecules, ions) of a substance.

Copyright © Sterling Education.

Ion – a molecule that has gained or lost electrons; an atom or a group of atoms carries an electric charge (Na^+).

Ion product for water – equilibrium constant for water ionization; $K\text{w} = [H_3O^+]\cdot[^-OH] = 1.00 \times 10^{-14}$ at 25 °C.

Ionic bond – the electrostatic attraction between oppositely charged ions, resulting from a transfer of electrons.

Ionic bonding – chemical bonding resulting from transferring electrons from one atom or group.

Ionic compounds – compounds containing predominantly ionic bonding.

Ionic geometry – arrangement of atoms (not lone pairs of electrons) about the central atom of a polyatomic ion.

Ionization – the breaking up of a compound into separate ions; in an aqueous solution, the process by which a molecular compound reacts with water and forms ions.

Ionization constant – equilibrium constant for the ionization of a weak electrolyte.

Ionization energy – the minimum amount of energy required to remove the most loosely held electron of an isolated gaseous atom or ion.

Ion exchange – a method of removing hardness from water, it replaces the positive ions that cause the hardness with H^+ ions.

Ionization isomers – result from the interchange of ions inside and outside the coordination sphere.

Isoelectric – having the same electronic configurations.

Isomers – different substances with the same formula.

Isomorphous – refers to crystals having the same atomic arrangement.

Isotopes – two or more forms of atoms of the same element with different masses; atoms containing the same number of protons but different numbers of neutrons.

IUPAC – acronym for "International Union of Pure and Applied Chemistry."

J

Joule (J) – a unit of energy in the SI system; one joule is 1 $kg\cdot m^2/s^2$, which is 0.2390 calories.

K

K capture – absorption of a *K* shell (n = 1) electron by a proton as it is converted to a neutron.

Kelvin – a unit of measure for temperature based upon an absolute scale.

Kinetics – a subfield of chemistry specializing in reaction rates.

Kinetic energy (*KE*) – energy that matter processes through its motion.

Kinetic Molecular Theory – a theory that attempts to explain macroscopic observations on gases in microscopic or molecular terms.

Any duplication (copies, uploads, PDFs) is illegal.

L

Lanthanides – elements 57 (lanthanum) through 71 (lutetium); grouped because of their similar behavior in chemical reactions.

Lanthanide contraction – a decrease in the radii of the elements following the lanthanides compared to what would be expected if there were no *f*-transition metals.

Latent heat – the energy absorbed or released when a substance changes state without changing temperature.

Lattice – a unique arrangement of atoms or molecules in a crystalline liquid or solid.

Law of combining volumes (Gay-Lussac's Law) – at constant temperature and pressure, the volumes of reacting gases (and any gaseous products) can be expressed as ratios of small whole numbers.

Law of conservation of energy – energy cannot be created or destroyed; it can only change form.

Law of conservation of matter – there is no detectable change in the quantity of matter during an ordinary chemical reaction.

Law of conservation of matter and energy – the amount of matter and energy in the universe is fixed.

Law of definite proportions (law of constant composition) – different samples of a pure compound contain the same elements in the same proportions by mass.

Law of partial pressures (or *Dalton's Law*) – the pressure exerted by a mixture of gases is the sum of the partial pressures of the individual gases.

Laws of thermodynamics – physical laws which define quantities of thermodynamic systems describe how they behave and (by extension) set certain limitations, such as perpetual motion.

Lead storage battery – a secondary voltaic cell used in most automobiles.

Leclanche cell – a common type of *dry cell.*

Le Chatelier's principle – states that a system at equilibrium, or striving to attain equilibrium, responds in such a way as to counteract any stress placed upon it; if stress (change of conditions) is applied to a system at equilibrium, the system will shift in the direction that reduces stress.

Leveling effect – acids stronger than the acid characteristic of the solvent react with the solvent to produce that acid; a similar statement applies to bases. The strongest acid (base) that can exist in a given solvent is the acid (base) characteristic of the solvent.

Levorotatory (or *levo*) – an optically active substance rotates plane-polarized light counterclockwise.

Lewis acid – any species that can accept a share in an electron pair.

Lewis base – any species that can make available a pair of electrons.

Lewis dot formula (electron dot formula) – representation of a molecule, ion, or formula unit by showing atomic symbols and only outer shell electrons.

Ligand – a Lewis base in a coordination compound.

Light – that portion of the electromagnetic spectrum visible to the naked eye; known as "visible light."

Limiting reactant – a substance that stoichiometrically limits the number of products formed.

Copyright © Sterling Education.

Linear accelerator – a device used for accelerating charged particles along a straight line path.

Line spectrum – an atomic emission or absorption spectrum.

Linkage isomers – a particular ligand bonds to a metal ion through different donor atoms.

Liquid – a state of matter which takes the shape of its container.

Liquid aerosol – colloidal suspension of a liquid in a gas.

London dispersion forces – very weak and very short-range attractive forces between short-lived temporary (induced) dipoles; known as "dispersion forces."

Lone pair – a pair of electrons residing on one atom and not shared by other atoms; an unshared pair.

Low spin complex – crystal field designation for an inner orbital complex; contains electrons paired t_{2g} orbitals before e_g orbitals are occupied in octahedral complexes.

Lubricant – a substance capable of reducing friction (i.e., a force that opposes the direction of motion).

M

Magnetic field – a space around a magnet where magnetism can be detected.

Magnetic quantum number – quantum mechanical solution to a wave equation designating the orbital within a given set (*s*, *p*, *d*, *f*) in which an electron resides.

Manometer – a two-armed barometer.

Mass (m) – a measure of the amount of matter in an object; mass is usually measured in grams or kilograms.

Mass action expression – for a reversible reaction, aA + bB cC + dD; the product of the concentrations of the products (species on the right), each raised to the power corresponding to its coefficient in the balanced chemical equation, divided by the product of the concentrations of reactants (species on the left), each raised to the power corresponding to its coefficient in the balanced equation; at equilibrium the mass action expression is equal to K.

Mass deficiency – the amount of matter converted into energy when an atom forms from constituent particles.

Mass number – the sum of the numbers of protons and neutrons in an atom; an integer.

Mass spectrometer – an instrument that measures the charge-to-mass ratio of charged particles.

Matter – anything that has mass and occupies space.

Mechanism – the sequence of steps by which reactants are converted into products.

Melting point – the temperature at which liquid and solid coexist in equilibrium.

Meniscus – the shape assumed by the surface of a liquid in a cylindrical container.

Melting – the phase change from a solid to a liquid.

Metal – a chemical element that is a good conductor of electricity and heat and forms cations and ionic bonds with nonmetals; elements below and to the left of the stepwise division (metalloids) in the upper right corner of the periodic table; about 80% of known elements are metals.

Metallic bonding – bonding within metals due to the electrical attraction of positively charged metal ions for mobile electrons that belong to the crystal.

Metallic conduction – conduction of electrical current through a metal or along a metallic surface.

Metalloid – a substance with the properties of metals and nonmetals (B, Al, Si, Ge, As, Sb, Te, Po and At).

Metathesis reactions – reactions in which two compounds react to form two new compounds, with no changes in oxidation number; reactions in which the ions of two compounds exchange partners.

Method of initial rates – method of determining the rate-law expression by carrying out a reaction with different initial concentrations and analyzing the resultant changes in initial rates.

Methylene blue – a heterocyclic aromatic chemical compound with the molecular formula $C_{16}H_{18}N_3SCl$.

Miscible liquids – mix to form a solution (e.g., alcohol and water).

Miscibility – the ability of one liquid to mix with (dissolve in) another liquid.

Mixture – two or more different substances mingled together but not chemically combined. A sample of matter composed of two or more substances, each of which retains its identity and properties.

Moderator – a substance (e.g., deuterium, oxygen, paraffin) that slows fast neutrons upon collision.

Molality – a concentration expressed as the number of moles of solute per kilogram of solvent.

Molarity – the number of moles of solute per liter of solution.

Molar solubility – the number of moles of a solute that dissolves to produce a liter of a saturated solution.

Mole – a measurement of an amount of substance; a single mole contains approximately 6.022×10^{23} units or entities; abbreviated mol.

Molecule – a chemically bonded number of electrically neutral atoms.

Molecular equation – a chemical reaction in which formulas are written as if substances existed as molecules; only complete formulas are used.

Molecular formula – indicates the number of atoms present in a molecule of a molecular substance.

Molecular geometry – the arrangement of atoms (not lone pairs of electrons) around a central atom of a molecule or polyatomic ion.

Molecular orbital (*MO*) – resulting from the overlap and mixing of atomic orbitals on different atoms (i.e., a region where an electron can be found in a molecule, as opposed to an atom); an MO belongs to the molecule.

Molecular orbital theory – a theory of chemical bonding based upon postulated molecular orbitals.

Molecular weight – the mass of one molecule of a nonionic substance in atomic mass units.

Molecule – the smallest particle of a compound capable of stable, independent existence.

 Copyright © Sterling Education.

Mole fraction – the number of moles of a component in a mixture divided by the number of moles in the mixture.

Monoprotic acid – can form only one hydronium ion per molecule; may be strong or weak.

Mother nuclide – nuclide that undergoes nuclear decay.

N

Native state – refers to the occurrence of an element in an uncombined or free state in nature.

Natural radioactivity – spontaneous decomposition of an atom.

Neat – conditions with a liquid reagent or gas performed with no added solvent or co-solvent.

Nernst equation – corrects standard electrode potentials for nonstandard conditions.

Net ionic equation – results from canceling spectator ions and eliminating brackets from a total ionic equation.

Neutralization – the reaction of an acid with a base to form a salt and water; usually, the reaction of hydrogen ions with hydroxide ions to form water molecules.

Neutrino – a particle that travels at speeds close to the speed of light; created due to radioactive decay.

Neutron – a neutral unit or subatomic particle with no net charge and a mass of 1.0087 amu.

Nickel-cadmium cell (NiCd battery) – a dry cell where the anode is Cd, the cathode is NiO_2, and the electrolyte is basic.

Nitrogen cycle – the complex series of reactions by which nitrogen is slowly but continually recycled in the atmosphere, lithosphere, and hydrosphere.

Noble gases – elements of the periodic Group 0; He, Ne, Ar, Kr, Xe, Rn; known as "rare gases;" formerly called "inert gases."

Nodal plane – a region in which the probability of finding an electron is zero.

Nonbonding orbital – a molecular orbital derived only from an atomic orbital of one atom; it lends neither stability nor instability to a molecule or ion when populated with electrons.

Nonelectrolyte – a substance whose aqueous solutions do not conduct electricity.

Nonmetal – an element that is not metallic.

Nonpolar bond – a covalent bond in which electron density is symmetrically distributed.

Nuclear – of or about the atomic nucleus.

Nuclear binding energy – the energy equivalent of the mass deficiency; the energy released in forming an atom from the subatomic particles.

Nuclear fission – when a heavy nucleus splits into nuclei of intermediate masses, and protons are emitted.

Nuclear magnetic resonance spectroscopy – a technique that exploits the magnetic properties of specific nuclei; helps identify unknown compounds.

Any duplication (copies, uploads, PDFs) is illegal.

Nuclear reaction – involves a change in the composition of a nucleus and can emit or absorb a tremendous amount of energy.

Nuclear reactor – a system in which controlled nuclear fission reactions generate heat energy on a large scale, subsequently converted into electrical energy.

Nucleons – particles comprising the nucleus: protons and neutrons.

Nucleus – the tiny and dense, positively charged center of an atom containing protons and neutrons, as well as other subatomic particles; the net charge is positive.

Nuclides – refers to different atomic forms of elements; in contrast to isotopes, which refer only to different atomic forms of a single element.

Nuclide symbol – designation for an atom A/Z E, in which E is the symbol of an element, Z is its atomic number, and A is its mass number.

Number density – a measure of the concentration of countable objects (e.g., atoms, molecules) in space; the number per volume.

O

Octahedral – molecules and polyatomic ions with one atom in the center and six atoms at the corners of an octahedron.

Octane number – a number that indicates how smoothly a gasoline burns.

Octet rule – during bonding, atoms tend to reach an electron arrangement with eight electrons in the outermost shell. Many representative elements attain at least a share of eight electrons in their valence shells when they form molecular or ionic compounds; there are some limitations.

Open sextet – species with only six electrons in the highest energy level of the central element (many Lewis acids).

Orbital – may refer to an atomic orbital or a molecular orbital.

Organic chemistry – the chemistry of substances that contain carbon-hydrogen bonds.

Organic compound – substances that contain carbon.

Osmosis – when solvent molecules pass through a semi-permeable membrane from a dilute solution into a more concentrated solution.

Osmotic pressure – the hydrostatic pressure produced on the surface of a semi-permeable membrane.

Outer orbital complex – valence bond designation for a complex in which the metal ion utilizes d orbitals in the outermost (occupied) shell in hybridization.

Overlap – the interaction of orbitals on different atoms in the same region of space.

Oxidation – the addition of oxygen or the loss of electrons. An algebraic increase in the oxidation number may correspond to a loss of electrons.

Oxidation numbers – quantitative values used as mechanical aids in writing formulas and balancing equations; for single-atom ions, they correspond to the charge on the ion; more electronegative atoms are assigned negative oxidation numbers, known as *oxidation states*.

Copyright © Sterling Education.

Oxidation-reduction reactions – reactions in which oxidation and reduction occur; known as *redox reactions*.

Oxide – a binary compound of oxygen.

Oxidizing agent – the substance that oxidizes another substance and is reduced.

P

Pairing – a favorable interaction of two electrons with opposite m values in the same orbital.

Pairing energy – the energy required to pair two electrons in the same orbital.

Paramagnetism – attraction toward a magnetic field, stronger than diamagnetism but still weak compared to ferromagnetism.

Partial pressure – the force exerted by one gas in a mixture of gases.

Particulate matter – fine, divided solid particles suspended in polluted air.

Pauli exclusion principle – no electrons in the same atom may have identical sets of four quantum numbers.

Percentage ionization – the percentage of the weak electrolyte that will ionize in a solution of given concentration.

Percent by mass – 100% times the actual yield divided by the theoretical yield.

Percent composition – the mass percent of each element in a compound.

Percent purity – the percent of a specified compound or element in an impure sample.

Period – the elements in a horizontal row of the periodic table.

Periodicity – regular periodic variations of properties of elements with their atomic number (and position in the periodic table).

Periodic Law – the properties of the elements are periodic functions of their atomic numbers.

Periodic table – an arrangement of elements by increasing atomic numbers, emphasizing periodicity.

Peroxide – a compound with oxygen in –1 oxidation state; metal peroxides contain the peroxide ion, O_2^{2-}.

pH – the measure of acidity (or basicity) of a solution; negative logarithm of the concentration (mol/L) of the H_3O^+ [H^+] ion; scale is commonly used over a range 0 to 14.

Phase diagram – shows the equilibrium temperature-pressure relationships for different phases of a substance.

pH scale – a range from 0 to 14. If the pH of a solution is 7, it is neutral; if the pH of a solution is less than 7, it is acidic; if the pH of a solution is greater than 7, it is basic.

Permanent hardness – hardness (relative to lathering soap) in water that cannot be removed by boiling; caused by calcium sulfate.

Any duplication (copies, uploads, PDFs) is illegal.

Photoelectric effect – emission of an electron from the surface of a metal caused by impinging electromagnetic radiation of specific minimum energy; the current increases with increasing radiation intensity.

Photon – a carrier of electromagnetic radiation of all wavelengths, such as gamma rays and radio waves; known as a *quantum of light*.

Physical change – when a substance changes from one physical state to another, but no substances with different compositions are formed; physical change may involve a phase change (e.g., melting, freezing) or another physical change, such as crushing a crystal or separating one volume of liquid into different containers; it does not produce a new substance.

Plasma – a physical state of matter that exists at extremely high temperatures in which molecules are dissociated, and most atoms are ionized.

Polar bond – a covalent bond with an unsymmetrical distribution of electron density.

Polarimeter – a device used to measure optical activity.

Polarization – the buildup of a product of oxidation or reduction of an electrode, preventing further reaction.

Polydentate – refers to ligands with more than one donor atom.

Polyene – a compound that contains more than one double bond per molecule.

Polymerization – the combination of many small molecules to form large molecules.

Polymer – a large molecule consisting of chains or rings of linked monomer units, usually characterized by high melting and boiling points.

Polymorphous – refers to substances that can crystallize in more than one crystalline arrangement.

Polyprotic acid – forms two or more H_3O^+ ions per molecule; often, at least one ionization step is weak.

Positron – a nuclear particle with the mass of an electron but opposite charge (positive).

Potential difference (or *voltage*) – the force that moves the electrons around circuit; unit is Volt (V).

Potential energy (*PE*) – energy stored in a body or a system due to its position in a force field or configuration.

Power – the rate at which energy is converted from one form to another; the unit is Watts (W). Power = voltage × current (P = VI).

Precipitate – an insoluble solid formed by mixing in solution the constituent ions of a slightly soluble solution.

Precision – how close the results of multiple experimental trials are; see *accuracy*.

Pressure – force per unit area; unit is Pascal (Pa).

Primary standard – a known high degree of purity substance that undergoes one invariable reaction with the other reactant of interest.

Primary voltaic cells – voltaic cells that cannot be recharged; no further chemical reaction is possible once the reactants are consumed.

Copyright © Sterling Education.

Products – chemicals produced (from reactants) in a chemical reaction.

Proton – a subatomic particle having a mass of 1.0073 amu and a charge of +1, found atom's nucleus.

Protonation – the addition of a proton (H^+) to an atom, molecule, or ion.

Pseudobinaryionic compounds – contain more than two elements but are named like binary compounds.

Q

Quanta – the minimum amount of energy emitted by radiation.

Quantum mechanics – the study of how atoms, molecules, subatomic particles, etc., behave and are structured; a mathematical method of treating particles based on quantum theory, which assumes that energy (of small particles) is not infinitely divisible.

Quantum numbers – numbers that describe the energies of electrons in atoms; derived from quantum mechanical treatment.

Quarks – elementary particles and a fundamental constituent of matter, combining to form hadrons (i.e., protons and neutrons).

R

Radiation – 1) heat transfer through invisible rays, which travel outwards from the hot object without a medium. 2) high-energy particles or rays emitted during the nuclear decay processes.

Radical – an atom or group of atoms that contains one or more unpaired electrons; usually a very reactive species.

Radioactive dating – a method of dating ancient objects by determining the ratio of mother and daughter nuclides present in an object and relating the ratio to the object's age via half-life calculations.

Radioactive tracer – a small amount of radioisotope replacing a nonradioactive isotope of the element in a compound whose path (e.g., in the body) or whose decomposition products are monitored by detection of radioactivity; known as a "radioactive label."

Radioactivity – the spontaneous disintegration of atomic nuclei.

Raoult's Law – the vapor pressure of a solvent in an ideal solution decreases as its mole fraction decreases.

Rate-determining step – the slowest step in a mechanism; determines the overall reaction rate.

Rate-law expression – an equation relating the reaction rate to the concentrations of the reactants and the specific rate of the reaction.

Rate of reaction – the change in the concentration of a reactant or product per unit time.

Reactants – substances consumed in a chemical reaction; react together in a chemical reaction.

Reaction quotient – the mass action expression under any set of conditions (not necessarily equilibrium); its magnitude relative to K determines the direction in which the reaction must occur to establish equilibrium.

Any duplication (copies, uploads, PDFs) is illegal.

Reaction ratio – the relative amounts of reactants and products involved in a reaction; may be the ratio of moles, millimoles, or masses.

Reaction stoichiometry – describes the quantitative relationships among substances participating in chemical reactions.

Reactivity series (or activity series) – an empirical, calculated, and structurally analytical progression of a series of metals, arranged by "reactivity" from highest to lowest; used to summarize information about the reactions of metals with acids and water, double displacement reactions, and the extraction of metals from ores.

Reagent – a substance (or compound) added to a system to cause a chemical reaction or to visualize if a reaction occurs; the terms reactant and reagent are often used interchangeably; however, a *reactant* is more specifically a substance consumed during a chemical reaction.

Reducing agent – a substance that reduces another substance and is itself oxidized.

Reduction – the removal of oxygen or the gaining of electrons.

Resonance – the concept in which two or more equivalent dot formulas for the same arrangement of atoms (resonance structures) are necessary to describe the bonding in a molecule or ion.

Reverse osmosis – forcing solvent molecules to flow through a semi-permeable membrane from a concentrated solution into a dilute solution by applying greater hydrostatic pressure on the concentrated side than the osmotic pressure opposing it.

Reversible reaction – processes that do not go to completion and occur in the forward and reverse direction.

S

Saline solution – a general term for NaCl (i.e., sodium chloride) in water.

Salt – when a metal replaces the hydrogen of an acid.

Salts – ionic compounds composed of anions and cations.

Salt bridge – a U-shaped tube containing an electrolyte, connects the two half-cells of a voltaic cell.

Saturated solution –no more solute will dissolve at that temperature.

***s*-block elements** – group 1 and 2 elements (alkali and alkaline metals), including hydrogen and helium.

Schrödinger equation – quantum state equation representing the behavior of an electron around an atom; describes the wave function of a physical system evolving.

Second Law of Thermodynamics – the universe tends toward a state of greater disorder in spontaneous processes.

Secondary standard – a solution that has been titrated against a primary standard; a standard solution.

Secondary voltaic cells – voltaic cells that can be recharged; original reactants can be regenerated by reversing the direction of the current flow.

Semiconductor – a substance that does not conduct electricity at low temperatures but will do so at higher temperatures.

Copyright © Sterling Education.

Semi-permeable membrane – a thin partition between two solutions through which specific molecules can pass but others cannot.

Shielding effect – electrons in filled sets of *s*, *p* orbitals between the nucleus and outer shell electrons shield the outer shell electrons somewhat from the effect of protons in the nucleus; known as the "screening effect."

Sigma (σ) bonds – bonds resulting from the head-on overlap of atomic orbitals. The region of electron sharing is along and (cylindrically) symmetrical to the imaginary line connecting the bonded atoms.

Sigma orbital – molecular orbital resulting from the head-on overlap of two atomic orbitals.

Single bond – covalent bond resulting from the sharing of two electrons (one pair) between two atoms.

Sol – a suspension of solid particles in a liquid; artificial examples include sol-gels.

Solid – one of the states of matter, where the molecules are packed closely, resistance to movement/deformation, and volume change.

Solubility product constant – equilibrium constant for the dissolution of a slightly soluble compound.

Solubility product principle – the solubility product constant expression for a slightly soluble compound is the product of the concentrations of the constituent ions; each raised to the power that corresponds to the number of ions in one formula unit.

Solute – the dispersed (i.e., dissolved) phase of a solution; the solution is mixed into the solvent (e.g., NaCl in saline water).

Solution – a homogeneous mixture of multiple substances; comprised of solutes and solvents; a mixture of a solute (usually a solid) and a solvent (usually a liquid).

Solvation – the process by which solvent molecules surround and interact with solute ions or molecules.

Solvent – the dispersing medium of a solution (e.g., H_2O in saline water).

Solvolysis – the reaction of a substance with the solvent in which it is dissolved.

***s*-orbital** – a spherically symmetrical atomic orbital; one per energy level.

Specific gravity – the ratio of the density of a substance to the density of water.

Specific heat – the amount of heat required to raise the temperature of one gram of substance 1 °C.

Specific rate constant – an experimentally determined (proportionality) constant; different for different reactions, and which changes only with temperature; *k* in the rate-law expression: Rate = k [A] × [B].

Spectator ions – ions in a solution that do not participate in a chemical reaction.

Spectral line – any of several lines corresponding to definite wavelengths of an atomic emission or absorption spectrum; marks the energy difference between two energy levels.

Spectrochemical series – arrangement of ligands in order of increasing ligand field strength.

Spectroscopy – the study of radiation and matter, such as X-ray absorption and emission spectroscopy.

Spectrum – display of component wavelengths (colors) of electromagnetic radiation.

Speed of light – the speed at which radiation travels through a vacuum (299,792,458 m/sec).

Any duplication (copies, uploads, PDFs) is illegal.

Square planar – describes molecules and polyatomic ions with one atom in the center and four atoms at the corners of a square.

Square planar complex – relationship with metal in the center of a square plane, with ligand donor atoms at each of the four corners.

Standard conditions for temperature and pressure (STP) – used to compare experimental results (25 °C and 100.000 kPa).

Standard electrodes – half-cells in which the oxidized and reduced forms of a species are present at the unit activity (1.0 M solutions of dissolved ions, 1.0 atm partial pressure of gases, pure solids, and liquids).

Standard electrode potential – by convention, the potential (Eo) of a half-reaction as a reduction relative to the standard hydrogen electrode when species are present at unit activity.

Standard entropy – the absolute entropy of a substance in its standard state at 298 K.

Standard molar enthalpy of formation – the amount of heat absorbed in forming one mole of a substance in a specified state from its elements in their standard states.

Standard molar volume – space occupied by 1 mole of ideal gas under standard conditions; 22.4 liters.

Standard reaction – a process where the numbers of moles of reactants in the balanced equation, in their standard states, are entirely converted to the numbers of moles of products in the balanced equation, at their standard state.

State of matter – a homogeneous, macroscopic phase (e.g., gas, plasma, liquid, solid) in increasing concentration.

Stoichiometry – quantitative relationships of elements and compounds undergoing chemical changes.

Strong electrolyte – a substance that conducts electricity well in a dilute aqueous solution.

Strong field ligand – a ligand that exerts a strong crystal or ligand electrical field and generally forms low-spin complexes with metal ions when possible.

Structural isomers – compounds that contain the same number and kinds of atoms, but with different geometries.

Subatomic particles – comprise an atom (e.g., protons, neutrons, electrons).

Sublimation – the direct vaporization of a solid by heating without passing through the liquid state; a phase transition from solid to gas.

Substance – any matter, specimens with the same chemical composition and physical properties.

Substitution reaction – a reaction in which another atom or group of atoms replaces an atom or a group of atoms.

Supercooled liquids – liquids that, when cooled, apparently solidify but continue to flow very slowly under the influence of gravity.

Supercritical fluid – a substance at a temperature above its critical temperature.

Supersaturated solution – contains a higher than saturation concentration of solute; slight disturbance or seeding causes crystallization of excess solute.

Copyright © Sterling Education.

Suspension – a heterogeneous mixture in which solute-like particles settle out of the solvent-like phase sometime after their introduction. A mixture of a liquid and a finely divided insoluble solid.

T

Talc – a mineral representing the Mohs Scale, composed of hydrated magnesium silicate with the chemical formula $H_2Mg_3(SiO_3)_4$ or $Mg_3Si_4O_{10}(OH)_2$.

Temperature – a measure of heat intensity (i.e., hotness or coldness of a sample); a measure of kinetic energy of an object. Units are measured in degrees, and scales include Celsius, Fahrenheit, and Kelvin.

Temporary hardness – hardness in water, removed by boiling; caused by calcium hydrogen carbonate.

Ternary acid – a ternary compound containing H, O and another element, often a nonmetal.

Ternary compound – a compound consisting of three elements; may be ionic or covalent.

Tetrahedral – a term used to describe molecules and polyatomic ions with one atom in the center and four atoms at the corners of a tetrahedron.

Theoretical yield – the maximum amount of a specified product that could be obtained from specified amounts of reactants, assuming complete consumption of the limiting reactant according to only one reaction and complete recovery of the product; see *actual yield.*

Theory – a model describing the nature of a phenomenon.

Thermal conductivity – a property of a material to conduct heat (often noted as k).

Thermal cracking – decomposition by heating a substance in the presence of a catalyst and without air.

Thermochemistry – the study of absorption/release of heat within a chemical reaction; studies heat energy associated with chemical reactions and physical transformations.

Thermodynamics – studying the effects of changing temperature, volume, or pressure (or work, heat, and energy) on a macroscopic scale.

Thermodynamic stability – when a system is in its lowest energy state with its environment (equilibrium).

Thermometer – a device that measures the average energy of a system.

Thermonuclear energy – energy from nuclear fusion reactions.

Third Law of Thermodynamics – entropy of a pure crystalline substance at absolute zero equals zero.

Titration – a procedure in which one solution is added to another solution until the chemical reaction between the two solutions is complete; the concentration of one solution is known, and that of the other is unknown. The process of adding one solution to a measured amount of another to find out exactly how much of each is required to react.

Torr – a unit to measure pressure; 1 Torr is equivalent to 133.322 Pa or 1.3158×10^{-3} atm.

Total ionic equation – the expression for a chemical reaction written to show the predominant form of species in aqueous solution or contact with water.

Transition elements (metals) – B Group elements except IIB in the periodic table; sometimes called transition elements, elements with incomplete *d* sub-shells; the *d*-block elements.

Any duplication (copies, uploads, PDFs) is illegal.

Transition state theory –reactants pass through high-energy transition states before forming products.

Transuranic element – an atomic number greater than 92; none of the transuranic elements are stable.

Triple bond – the sharing of three pairs of electrons within a covalent bond (e.g., N_2).

Triple point – where the temperature and pressure of three phases are the same; water has a unique phase diagram.

Tyndall effect – results from light scattering by colloidal particles (a mixture where one substance is dispersed evenly throughout another) or by suspended particles.

U

Uncertainty –any measurement that involves estimating any amount that cannot be precisely reproduced.

Uncertainty principle – knowing the location of a particle makes the momentum uncertain, while knowing the momentum of a particle makes the location uncertain.

Unit cell – the smallest repeating unit of a lattice.

Unit factor – statements used in converting between units.

Universal (or ideal) gas constant – proportionality constant in the ideal gas law (0.08206 L·atm/(K·mol)).

UN number – a four-digit code used to note hazardous and flammable substances.

Unsaturated hydrocarbons – hydrocarbons that contain double or triple carbon-carbon bonds.

V

Valence bond theory – proposes that covalent bonds are formed when atomic orbitals on different atoms overlap and the electrons are shared.

Valence electrons – outermost electrons of atoms; usually those involved in bonding.

Valence shell electron pair repulsion theory (VSEPR) – assumes electron pairs are arranged around the central element of a molecule or polyatomic ion with maximum separation (and minimum repulsion) among regions of high electron density.

Valency – the number of electrons an atom wants to gain, lose, or share to have a full outer shell.

Van der Waals' equation – a quantitative relationship of a state extending the ideal gas law to real gases by including two empirically determined parameters, which are specific for different gases.

Van der Waals force – one of the forces (attraction/repulsion) between molecules.

Van't Hoff factor – the ratio of moles of particles in solution to moles of solute dissolved.

Vapor – when a substance is below the critical temperature in the gas phase.

Vaporization – the phase change from liquid to gas.

Vapor pressure – the particle pressure of vapor at the surface of its parent liquid.

Copyright © Sterling Education.

Viscosity – the resistance of a liquid to flow (e.g., oil has a higher viscosity than water).

Volt – one joule of work per coulomb; the unit of electrical potential transferred.

Voltage – the potential difference between two electrodes; a measure of the chemical potential for a redox reaction.

Voltaic cells – electrochemical cells in which spontaneous chemical reactions produce electricity; known as *galvanic cells*.

Voltmeter – an instrument that measures the cell potential.

Volumetric analysis – measuring the volume of a solution (of known concentration) to determine the substance's concentration within the solution; see *titration*.

W

Water equivalent – the amount of water absorbing the same heat as the calorimeter per degree of temperature increase.

Weak electrolyte – a substance that conducts electricity poorly in a dilute aqueous solution.

Weak field ligand – a ligand that exerts a weak crystal or ligand field and generally forms high-spin complexes with metals.

X

X-ray – electromagnetic radiation between gamma and UV rays.

X-ray diffraction – a method for establishing structures of crystalline solids using single-wavelength X-rays and studying the diffraction pattern.

X-ray photoelectron spectroscopy – a spectroscopic technique used to measure the composition of a material.

Y

Yield – the amount of product produced during a chemical reaction.

Z

Zone melting – remove impurities from an element by melting and slowly traveling it down an ingot (cast).

Zone refining – a method of purifying a metal bar by passing it through an induction heater; this causes impurities to move along a melted portion.

Zwitterion (formerly called a dipolar ion) – a neutral molecule with a positive and negative electrical charge; multiple positive and negative charges can be present, distinct from dipoles at different locations within that molecule; known as *inner salts*.

Any duplication (copies, uploads, PDFs) is illegal.

Frank J. Addivinola, Ph.D.

This study guide's lead author and chief editor is Dr. Frank Addivinola. With his outstanding education, laboratory research, and decades of university science teaching, Dr. Addivinola lent his expertise to oversee the development of this series.

During his extensive career, Dr. Addivinola held faculty positions at colleges and universities, including Harvard University, Johns Hopkins University, University of Maryland, and Northeastern University, and taught undergraduate and graduate-level courses in biology, biochemistry, organic chemistry, inorganic chemistry, physics, anatomy and physiology, medical terminology, nutrition, and medical ethics. He received several awards for his research and presentations.

Dr. Frank Addivinola conducted original research in developmental biology as a doctoral candidate and pre-IRTA fellow in Molecular and Cell Biology at the National Institutes of Health (NIH). His dissertation advisor was Nobel laureate Marshall W. Nirenberg, Chief of the Biochemical Genetics Laboratory at the National Heart, Lung, and Blood Institute (NHLBI). Before NIH, Dr. Addivinola researched prostate cancer in the Cell Growth and Regulation Laboratory of Dr. Arthur Pardee at the Dana Farber Cancer Institute of Harvard Medical School.

Dr. Addivinola holds an undergraduate degree in biology from Williams College. He completed his Masters at Harvard University, Masters in Biotechnology at Johns Hopkins University, and five other graduate degrees at the University of Maryland University College, Suffolk University, and Northeastern University.

Essential Chemistry Self-Teaching Guides

Electronic Structure & Periodic Table

Chemical Bonding

States of Matter & Phase Equilibria

Stoichiometry

Solution Chemistry

Chemical Kinetics & Equilibrium

Acids & Bases

Chemical Thermodynamics

Electrochemistry

Visit our Amazon store

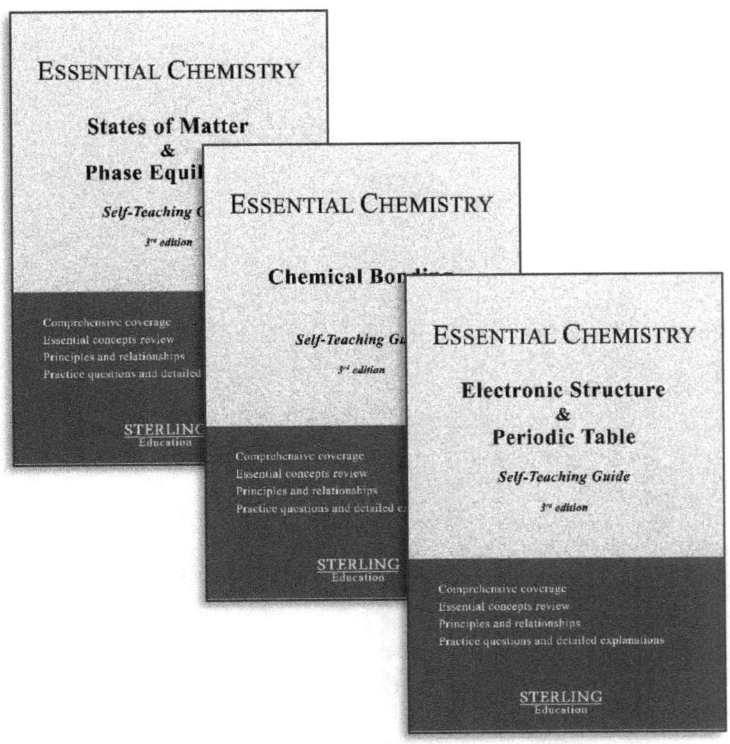

Essential Physics Self-Teaching Guides

Kinematics and Dynamics

Equilibrium and Momentum

Force, Motion and Gravitation

Work and Energy

Fluids and Solids

Waves and Periodic Motion

Light and Optics

Sound

Electrostatics and Electromagnetism

Electric Circuits

Thermal Physics

Atomic and Nuclear Physics

Visit our Amazon store

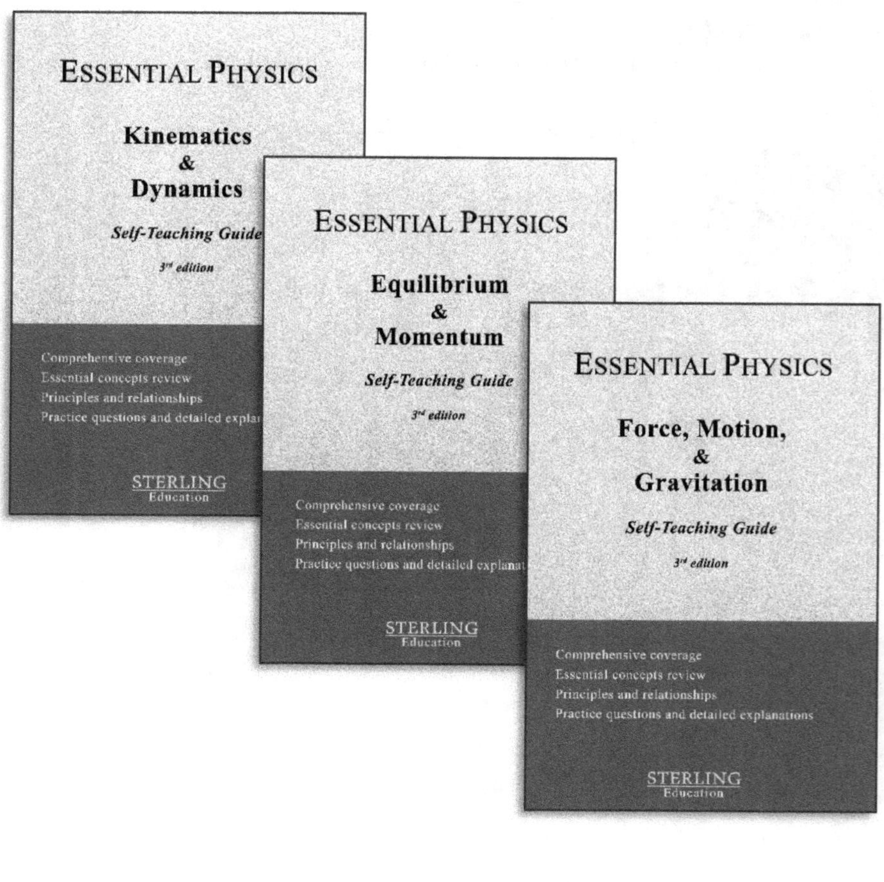

Essential Biology Self-Teaching Guides

Eukaryotic Cell & Cellular Metabolism

Molecular Biology & Genetics

Nervous & Endocrine Systems

Circulatory, Respiratory & Immune Systems

Digestive & Excretory Systems

Muscle, Skeletal & Integumentary Systems

Reproduction & Development

Microbiology

Plants & Photosynthesis

Evolution, Classification & Diversity

Ecology & Population Biology

Visit our Amazon store

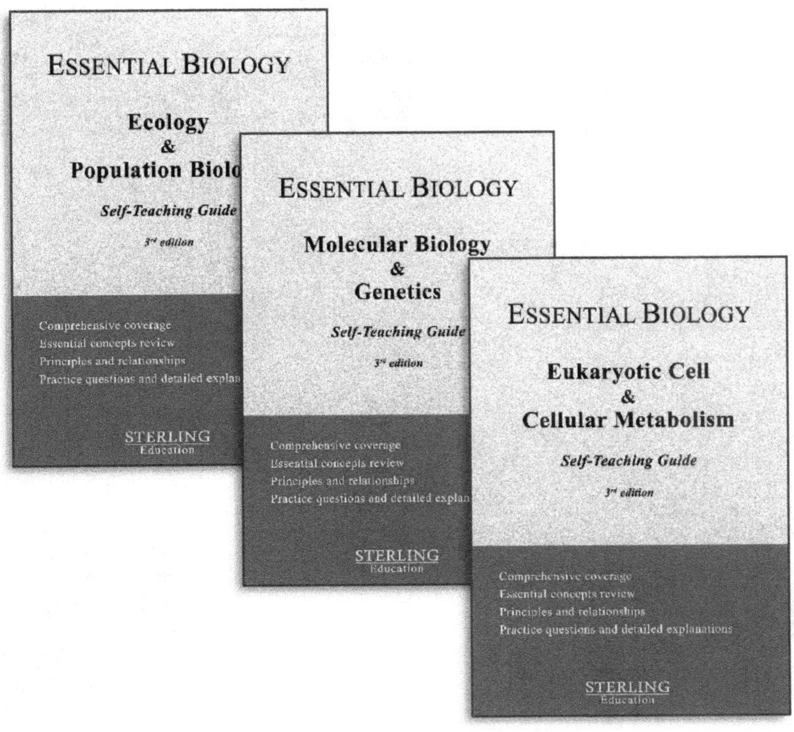

Everything You Always Wanted to Know About...

Chemistry

Physics

Cell & Molecular Biology

Organismal Biology

Human Anatomy & Physiology

Environmental Science

American History

American Law

American Government & Politics

Comparative Government & Politics

World History

European History

Psychology

Sociology

Human Geography

Visit our Amazon store

For online practice resources visit
https://www.sterling-prep.com

If you benefited from this book, please leave a review on Amazon so others can learn from your input. Reviews help us understand our customers' needs and experiences while keeping our commitment to quality.

www.ingramcontent.com/pod-product-compliance
Lightning Source LLC
LaVergne TN
LVHW082345200926
848481LV00012B/3457
* 9 7 9 8 8 8 5 5 7 1 1 1 1 *